THE
RED LIMIT

THE
RED LIMIT

The Search for the
Edge of the Universe

by TIMOTHY FERRIS

INTRODUCTION by CARL SAGAN

WILLIAM MORROW AND COMPANY, INC.
NEW YORK   1977

Library of Congress Cataloging in Publication Data

Ferris, Timothy.
The red limit.

Bibliography: p.
Includes index.
1. Astronomy—Popular works.   2. Cosmology.
I.   Title.
QB44.2.F47   523.1′09   76-54697
ISBN 0-688-03176-5

Book Design   Carl Weiss

FOR MY MOTHER AND FATHER,
AND IN MEMORY
OF MY BROTHER, BRUCE

PREFACE

We all dwell in a house of one room—the world with the
firmament for its roof—and are sailing the celestial spaces
without leaving any track.

—JOHN MUIR

IN THE TIME IT TAKES TO READ THIS SENTENCE, THE EARTH WILL
travel 200 miles in its orbit around the sun, the sun 3,000
miles in the rotation of our galaxy, and 350,000 miles more
space will open up between our galaxy and those of the
Hydra cluster, as the universe goes on expanding.

Who can comprehend such a thing, a million light-years
or a billion stars?

By some audacity, the source of which we do not
understand, we can write: *galaxy*. A scientist can write a
bundle of numbers that say *galaxy* and describe its size,
mass, composition and so forth. We can *refer to* something
the direct comprehension of which would probably rob us
of our sanity.

That process may be called abstraction. At the frontiers
of science it can lead us quite far from our experience;
there is little in everyday life to prepare us for the
spacetime continuum, the quantum principle, or the
notion of a finite but unbounded universe. Yet, after a

while, we can to get used to these odd concepts. As the physicist John Wheeler said about the quantum principle, paraphrasing Gertrude Stein on abstract art, "You look at it and it looks strange, and it looks strange, and it looks *very* strange, and then suddenly it doesn't look strange at all."

It cannot be denied that much is lost when we pretend to capture the wide world in scribbles on a page. The word *galaxy* is not a galaxy, *universe* is not the universe. The model of the natural world we build in our minds by such a process will forever be inadequate, just a little cathedral in the mountains. Still, it is better than no model at all. And to be building it, as the men and women on the pages that follow have helped build it, is to live a life of adventure.

Many physicists and astronomers gave generously of their time for interviews. Among them were Ralph Alpher, Halton Arp, John Bahcall, William Baum, Geoffrey and Margaret Burbidge, Robert Dicke, Thomas Gold, Jesse Greenstein, Robert Herman, Tom Kinman, Frank Low, Arno Penzias, Allan Sandage, Maarten Schmidt, C. Roger Lynds, John Wheeler and Robert Wilson.

I am deeply grateful to Bruce Partridge of Haverford College, who checked the manuscript for scientific accuracy and offered many excellent suggestions. His associate, Ralph Urban, made many valuable comments as well. Whatever errors may have persisted into print are, of course, entirely my responsibility.

Several of the quotations from John Wheeler appearing in Chapter 10 were taken from an interview conducted by Laurence B. Chase, who was kind enough to permit them to be reproduced here. David Allen of the Royal Greenwich Observatory graciously agreed to let me borrow some of his accounts of life at several of the world's major observatories.

The research was greatly aided by staff members of the Hayden Planetarium library and the libraries of Brooklyn College, New York City, Miami, Los Angeles and New York University.

Many friends offered encouragement or advice. Thanks especially to Cotton Broede, Nora Ephron, Karen Hitzig, Frederick Lunning, Thomas M. Powers, Paul Scanlon, Alex Shoumatoff, Erica Spellman, and Jack Thibeau.

My debt to Ann Druyan is inexpressible.

CONTENTS

INTRODUCTION

HUMAN BEINGS ARE, AT LEAST SO FAR, AN EXTRAORDINARILY successful species, dominating the land, sea, and air of their native planet, and now, in at least a preliminary way, setting forth to other places. The secret of our success is surely our curiosity, our intelligence, our manipulative abilities, and our passion for exploration—qualities that have been extracted painfully through billions of years of biological evolution. It is in the nature of mankind and the corollary of our success to ask and answer questions, and the deeper the question the more characteristically human is the activity. Today we are finding that a host of issues, once the exclusive province of philosophy and theology, are slowly yielding to scientific inquiry, that most human of human inventions. The structure of matter, the nature of consciousness, the origin and fundamentals of life, the motions of continents, the intelligence of other animals, the possibility of life on other planets, the formation and destinies of worlds, are all becoming accessible to the mind of man. This is not because we are today more intelligent than our ancestors of hundreds of thousands of years ago; but rather because our technology, the logical extension of our manipulative abilities, has at last reached a state of development adequate to probe these profound questions.

But the deepest of these questions is a subject thrilling even to contemplate: the origins, natures and fates not of worlds but of universes. As near as can be told, we have been preoccupied with this question as long as we have been human. It provides an essential part of the earliest body of myths and legends of cultures all over the planet

Earth. Many of these legends are, while beautiful, simplistic to the point of explaining nothing, and the world is imagined created out of the void by a primal being or a primal couple whose origin remains uncontemplated. The Greeks called such a being Chaos. In African and Asian myths there is a cosmic egg from which the world forms. Such ideas are the product of an extrapolation from everyday human experience to the cosmos at large. But there is no guarantee that everyday experience—on an insignificant dust mote in an immensity of space, and in an instant encompassed by the course of ages—may have any bearing whatever on cosmological issues. The depth of our understanding and the quality of our insights have been powerfully prefigured by the environment in which we have evolved. When we move our attention to areas in which we have no previous evolutionary experience—quantum mechanics, the world of the very small, say, or black holes, the world of the very dense—we find that the universe is not in accord with common-sense ideas. The universe is, of course, not obliged to conform to everyday notions on a small and obscure planet. Still, some early cosmological ideas are marked by a more impressive subtlety, as, for example, the occasional myths which posit an infinite regression of causes, or the view, in Book 10 of the *Rig Veda,* that even the gods may be ignorant of the origin of the universe.

We humans are far from being gods. And yet, in the decades of the twentieth century, we have been privileged to pierce at least a few of the veils that have clouded this subject. We have developed powerful new ways of thinking about the universe, largely connected with Einstein's theory of general relativity; and powerful methods of viewing the universe, chiefly large optical and sensitive radio telescopes. All cosmological questions are by no means answered. But a picture is emerging which in scale, scope and subtlety is one of the great triumphs of the human mind and hand.

We have accumulated not speculation but reasonably hard evidence on the extent, shape, age, contents and ultimate fate of the entire universe, and speculation has now expanded to include other possible universes, constructed on different principles and obeying different physical laws from those which we find apply to ours. With the further application of existing groundbased optical and radio telescopes, and particularly with the future launchings of large space telescopes, it is possible that almost all of the basic cosmological questions will be answered.

The ones that may long elude us are in some sense ultimate questions: why the physical laws are the way they are, where the matter of the universe came from "in the first place," and what was there before the universe existed. But these questions may not be real questions, operationally defined. They may be only apparent questions, tied to our use of words. If, for example, the universe is infinitely old, no question arises about its origin. If we can invent a source for the matter in the universe we immediately encounter the question of the origin of that source, and thereby come face to face with the standard problem of an infinite regression of causes. If we are trapped in a universe with only one set of physical laws, it is difficult to imagine experiments about which other categories of physical laws are possible. Much has been learned about cosmology recently; much will almost certainly be learned in the coming decade or two. But there is some comfort in the thought that we will never know everything. It would be a very dull universe for any intelligent being were everything of importance to be known.

Because the pace of recent discovery has been so rapid, and the subject matter so removed from everyday experience, the excitement and exhilaration and passion of modern cosmology is not appreciated nearly so well as it should be. There is a need for a comprehensible, accurate, up-to-date discussion of cosmology which does not talk down to the intelligent lay reader. I believe Timothy Ferris' *The Red Limit* is such a book. It is gracefully composed, studded with metaphors and similes of poetic clarity, and remarkably successful in conveying in words some of the content and feel of the mathematics essential to the subject. It has entrancing historical vignettes, many forgotten or generally unknown. It does not shy away from the vigorous controversy and strong personalities of the leading cosmologists of our age. It does not evade philosophical implications. It is not bashful about pronouncing mystery where we are still ignorant. For many readers who have not previously encountered modern cosmological ideas, this book will provide a twofold revelation—about the beauty and grandeur of the universe, and about the brilliance and tenacity of the human minds that occupy an obscure corner of that universe.

> — CARL SAGAN
> David Duncan Professor of Astronomy
>   and Space Sciences
> Director, Laboratory for Planetary
>   Studies
> Cornell University
> Ithaca, New York

When men lack a sense of awe, there will be disaster.

—Lao-Tzu (72)

1

THE EXPANSION OF THE UNIVERSE

Although the myriad things are many,

their order is one.

—CHUANG-TZU

NEW STARS BEING BORN AND OLD STARS DYING, SOLITARY
stars waltzing in pairs or trios or gathered in assemblies of
thousands, stars thin as a vacuum and stars hard as
diamond, red, white, blue and yellow stars, 100 billion stars
make up the congregation of the Milky Way galaxy. Our
sun, a yellow dwarf star orbited by nine planets, resides in
a spur of the Orion Arm, 30,000 light-years from the
galaxy's center and 20,000 light-years in from its outer
edge. Pristine streamers, glowing lagoons and tangled black
tracts of dust and gas are the stuff from which the galaxy
will, in time, fashion again as many more stars as it now
embraces. Announcing itself in a howl of electromagnetic
noise, magnetism and light as it glides through space and
time, the Milky Way measures 100,000 light-years across
and consumes 200 million years per stately spin. Its trailing
arms of dust, gas and stars, arrayed like the spiral convolu-
tions of a seashell, have carried the sun something under
one percent of one trip around the galactic center since
homo sapiens appeared on the earth.

The Milky Way is one galaxy of perhaps two dozen that
share membership in a cluster astronomers call the Local
Group. Our nearest neighbors are two satellite galaxies, the
Large and Small Magellanic Clouds, and a half dozen or so
dwarfs. Two million light-years away stands the Andromeda Galaxy, the dominant spiral of the group, an
outsized replica of the Milky Way. The population of the
Local Group is 500 billion stars.

Beyond, billions more galaxies recede in profusion in
deepening space. The antique light from many of them was

near completion of its long journey to our eyes when life on Earth had evolved so high as the sea urchin.

The universe contains at least as many galaxies as there are stars in the Milky Way.

Look up. A moonless late summer evening is a good time in the northern hemisphere. Face due east and you are looking out from the plane of our galaxy; out there, past a few thousand foreground stars, lies intergalactic space. Seven degrees above the star Beta Andromeda, you should be able to make out the soft glow of the Andromeda Galaxy. Turn and face due west, near the bright navigational star Arcturus. Here you are looking pretty well out of the whole Local Group. The inky sky between the foreground stars harbors clouds of remote galaxies.

From northeast to southwest the Milky Way arcs overhead, our galaxy viewed from within. Sweep it slowly with binoculars—the slower the better—from Cassiopeia in the northeast. You will see meadowlands of stars, cut with hints of glowing gas. One huge cloud of dark dust and gas reveals itself as a rift dividing the Milky Way from Cygnus to the southern horizon, like a rip in the sky. Approaching Sagittarius in the south, your field of view becomes heaped with stars. You are looking toward the heart of our galaxy.

For humans to have discovered that we live within a galaxy was a feat more prodigious, in terms of scale, than if, say, a band of protozoa in a Philippine tidal basin had charted the Pacific Ocean. It came about in a revelatory flurry early in this century, but was based on scrutiny of the skies going back three centuries.

When Galileo turned his small telescope to the night skies over seventeenth-century Padua, he found that the Milky Way, called by the Pythagoreans The Bridge Out of

Time, was composed of millions of stars. This was the first indication that we inhabit a vast but not limitless stellar system, one that has a characteristic flattened shape. In addition, Galileo found a number of small patches of glowing light, the nebulae, (from the Latin word for "mist"), whose diversity of size and shape hinted that understanding them might be complicated.

The eighteenth-century comet-hunter Charles Messier catalogued 103 nebulae, by way of warning other comet-seekers not to mistake them for legitimate prey. Many more were observed by the English astronomer William Herschel, who had a technologically premature passion for building big reflecting telescopes and who liked to boast, "I have looked farther into space than ever human being did before me." Herschel's son John continued his father's observations. Some of the nebulae looked like chalk-colored spiderweb tangled among the stars. Others were spiral in shape, resembling pinwheels. Through the nineteenth century, most people were inclined to think all nebulae were gas or dust in our own stellar system. The one remarkable exception was the philosopher Immanuel Kant, who perceived, with little but reasoned intuition to guide him, that the delicate pinwheel nebulae might be galaxies like our own.

In 1751 Kant, at that time a tutor in Königsberg, read a newspaper story about the speculative cosmologies of Thomas Wright, a pious English surveyor and amateur scientist who authored several theories of the cosmos. Some of Wright's models were mutually contradictory—he proposed variously that the Milky Way was spherical or flat like a grindstone, composed of stars like the sun or just an illusion—but the contradictions did not seem to bother him; he wrote his theories as kapellmeisters composed cantatas, as offerings to the greater glory of God. By a lucky accident, the newspaper account Kant read badly oversimplified Wright, giving Kant the impression that Wright saw the Milky Way as a thin disk composed of stars, which he did not but which in fact it is. The idea appealed to Kant, and after four years of study he published, at age thirty-one, a slim, anonymous book that struck close to the truth. Titled *General History of Nature and Theory of the Heavens,* it correctly proposed that while some nebulae, those clearly associated with stars, lay within our own Milky Way, others, the spirals or oval-shaped nebulae, were separate Milky Ways at enormous distances. This made Kant the first to guess the true nature of the spiral "nebulae." The book attracted little attention, in part because there was then no way to test Kant's theory. That required the spectroscope.

In 1802 the English physicist William Wollaston found that by placing a thin slit in front of a prism he could break sunlight down into component parts so sharply defined that dark lines appeared along the spectrum, signaling the absorption of characteristic frequencies of light by atoms of various elements in the sun's outer layers. The ability of a prism to dissect sunlight into colors had been known for a long time (Isaac Newton discovered that in 1666) but Wollaston's refinements turned spectroscopy from a diversion into a scientific tool. A Bavarian optician, Joseph von Fraunhofer, soon built a more advanced spectroscope and found that the sun's spectrum, from deep red to violet, was interrupted by hundreds of black lines like gaps between piano keys.

John Herschel had learned that each chemical element, when heated and its glow analyzed through one of the new laboratory spectroscopes, produced a characteristic spectrum all its own. The English chemist Robert Bunsen and the German physicist Gustav Kirchhoff in 1859 compared laboratory spectra with a solar spectrum and found telltale lines of hydrogen, iron, sodium, magnesium, nickel and calcium in the sun. A question once considered the epitome of the unknowable—what are the stars made of—could, using the spectroscope, now be answered.

The first astronomer to train a spectroscope on the stars was Sir William Huggins, a wealthy gentleman who maintained an observatory on the roof of his home in Tulse Hill, London. Educated as a chemist, Huggins built a spectroscope, attached it to his telescope and immersed himself in the most exciting work of his life. Each distant star obligingly revealed to his spectroscope the chemical elements it was made of. "Nearly every night's work was red-lettered by some discovery," Huggins wrote happily. After satiating himself on starlight, he turned in 1864 to the nebulae, with results that helped support Kant's hypothesis. He found that spectra of the nebulae were divided into two distant kinds: Some clearly were composed of gas, while others displayed spectra much like the sun's, suggesting they were made up of stars. The spiral nebulae Huggins examined all had sunlike spectra.

Two theories prevailed concerning the spirals. Some held, with Kant, that they were external systems of stars. The majority leaned toward the view that they were whirlpools of gas, relatively nearby, each in the process of forming a new star. The whirlpool model had been put forth in 1796 by Pierre-Simon Laplace, a mathematician famous for his elegant analysis of how the planets move in their orbits. Laplace's book on the subject, *Essay on the System of the World,* sold well, while Kant's was almost unknown, and Laplace's model had a certain grace, even though he was misguided in applying it to the spiral nebulae. Huggins' spectroscope might have cleared up the question right away, but instead the issue was confused by a cataclysm that had occurred two million years earlier.

In two million B.C., Earth time, a star near the center of the Andromeda Galaxy exploded with tremendous violence. Some stars dying in this fashion do so as novae, a few as the still more powerful supernovae; this was an especially violent supernova. In a matter of days the star disgorged itself of energy and collapsed into cinder. Out through surrounding space spread a shell of light several light-days thick. Within 60,000 years the light had cleared its native galaxy. In 100,000 years it swept through two satellite galaxies, M32 and NGC 205, that ride like pilot fish near the big spiral. For 2 million years, light from the supernovae sped across intergalactic space, reaching the outposts of our galaxy roughly when the pyramids of Egypt were being built. To the temporary misfortune of science, it intercepted Earth in 1885, just in time to confuse the debate over the nebulae. An astronomer named Ernst Hartwig noticed the new star on August 20.

The Laplace hypothesis seemed confirmed. Here was a "new star," just where Laplace anticipated one, in the center of a swirling "nebula." The star soon faded, but that could perhaps be explained by its settling down after a traumatic birth. Kant's correct perception that the Andromeda spiral was a galaxy was discredited.

The overriding problem in understanding nebulae and stars alike was that astronomers did not know the distances of most of the objects they observed. A jewellike star floating prettily in a telescope's field of view might be 100 light-years away or 100,000; so might the Andromeda spiral.

Distances to a few hundred nearby stars had been painstakingly worked out by the parallax method, which consists of allowing Earth's motion around the sun to alter our perspective, so that neighboring stars appear to shift slightly relative to those in the background. But this method worked well only up to a few tens of light-years away. Beyond that, astronomers knew where stars were in the sky but not where they were in space. Until the third dimension could be supplied, talk of whether there were such things as galaxies, and if so whether ours was shaped like a cake, a pinwheel or a smoke ring, remained only conjecture. Herschel, again ahead of his time, tried to construct a model of the Milky Way by counting stars in all directions and simply assuming them to be of about the same intrinsic brightness, but stars come in all sorts of intrinsic brightnesses, and the task defeated him. He ended up with a model of the Milky Way resembling a thick botched pancake. He went on to other matters.

Stuck in two dimensions, astronomers in the closing decades of the nineteenth century shelved most big questions and bent themselves to less inspiring work, compiling star charts and assembling huge catalogues that listed the names, spectra and color of stars. Late nineteenth-century university astronomy departments resembled factories of the period, where imperious bosses oversaw rooms full of employees whose repetitious toil threatened their eyesight. But their methods produced results. The columns of dry figures suggested secrets about the stars.

At Harvard College Observatory, the menial jobs were held mostly by women. Edward Charles Pickering hired them, gave them the title "computer," and paid them twenty-five cents an hour to fill blank catalogue pages with tiny black ink numbers, no mistakes permitted. A methodi-

cal man, Pickering came to Harvard from M. I. T. in 1876,
stayed until his death in 1919, and spent much of that time
photographing and indexing the stars. With the help of his
younger brother William he set up an observatory in the
clear air of Flagstaff, Arizona, and another in Arequipa,
Peru, to photograph the northern and southern skies
systematically with standardized films and exposure times.
Spectra for thousands of stars also were recorded at the two
stations. The exposed plates were sent to Cambridge. Over
the years Pickering's collection grew to a quarter of a
million photographs, 225,000 spectra, and data filling
seventy volumes of the Harvard Observatory *Annals.* About
forty women worked as computers for Pickering. Most
came and went silently, but a few persisted in trying to
make some sense out of the forest of data they had been
given to harvest.

One was Annie Jump Cannon. Pickering befriended her
and allowed her to sit in on his physics lectures at M. I. T.,
normally barred to women. While still an undergraduate
she established a women's physics laboratory, then took a
B. S. at Wellesley and went on to index Harvard sky plates
for Pickering without pay. By 1896, at age thirty-three, she
had become expert on the spectra of stars, and Pickering
gave her a small salary. She stayed at Harvard forty-two
years, discovering over 300 variable stars and indexing
300,000 spectra. Fourteen years of this won her Pickering's
personal accolade "astronomer," but, while she gained an
international reputation and received honorary degrees
from Oxford and three other universities, her name did not
appear in the Harvard catalogue. Her presence was finally
officially acknowledged by the university in 1938, when she
was seventy-five years old.

Another "computer," Henrietta Swan Leavitt, discovered what is called the period-luminosity relationship in Cepheid variable stars, and so gave astronomy its third dimension, making possible the discovery of galaxies and of the expansion of the universe. She came to Harvard from the Society for the Intercollegiate Instruction of Women (later Radcliffe) in 1895. Pickering stacked plates from the Peru station in front of her and told her to look for variable stars. Leavitt would study two photographs taken of the same area of sky at two different dates and check whether any of the thousands of stars had changed in apparent brightness, betraying itself as variable.

Variable stars had been observed for centuries and were known to fall into two categories, the eclipsing binaries and what might be called genuine variables. Eclipsing binaries are systems of two or more stars that happen to be oriented in space so that one star, seen from our viewpoint, periodically passes in front of the other. True variables are stars that pulsate; their intrinsic brightness varies with time. A few take a year or more per pulse, but others, less leisurely, vary over periods of from a few hours to about a month. These more rapid variables included the Cepheids, so called because the first one identified was in the constellation Cepheus, named after the mythological king of Ethiopia who fathered Andromeda. The Cepheid variable stars proved to be the distance indicators astronomers needed.

Henrietta Leavitt examined thousands of Cepheids on the Harvard plates—she discovered 2,400 herself—and as time passed she began to perceive a pattern emerging. The brighter the Cepheid variable star, the longer it took to go through a cycle of brightness. She was able to make this important discovery because, as it happened, many of the Cepheids she was assigned to study were in the Large and

Small Magellanic Clouds, the two satellite galaxies of the Milky Way, 150,000 light-years distant. Pickering and his staff of computers did not know what the Clouds were, but an important effect operated anyway: All the Cepheids in them were bound up together at roughly the same distance, like fireflies in a bottle far away, so that confusing differences in brightness caused by varying distance were suppressed. The inherent relationship of period to luminosity revealed itself. If the true brightness, called absolute magnitude, of a single Cepheid star could now be determined, distances all over the cosmos might be measured.

Finding the distance to any single Cepheid proved difficult because none was nearby. But several astronomers, by exploiting a statistical method involving the sun's drift among its fellow stars, managed to get a fair estimate of the distance to a few short-period Cepheids. Now, it appeared, the genuine distance of any Cepheid variable could be determined from knowing only its apparent brightness and the period of its variability. Astronomy had found its third dimension.

Enter Harlow Shapley, a young man with slicked-back black hair, steady dark eyes, and an appetite for fame. Born in Missouri, he became a newspaper reporter at age fifteen, banging out stories for the Chanute, Kansas, *Daily Sun,* then the Joplin, Missouri, *Times* (where an editor once sent him to debunk a horse that could count, not in order to print an exposé but to blackmail the carnival into advertising). Shapley felt formal training might bring him loftier assignments, and in 1907 he arrived at the University of Missouri to study journalism, only to find the journalism school would not open for another year. Determined to study something, he chose astronomy. (Years

He then extended his distance deeper into space by developing a technique that became the basis for some of the most important work of twentieth-century astronomy. In each of the nearby clusters he isolated the most luminous stars, red giants and supergiants, and systematically compared their apparent brightness with that of the Cepheids. When he had done this often enough he felt he had a pretty good idea of the inherent brightness (called absolute magnitude) of the giant stars. Then he abandoned the relatively faint Cepheid variables, and used the giants as his beacons, or "standard candles," to get distances to more remote globular clusters where he could not identify Cepheids.

This work came in a storm. Of twenty papers published in the 1918–19 issues of the Mt. Wilson *Contributions,* half were by Shapley. Combining mountains of fresh data from the 60-inch telescope with Shapley's bold, rigorous reasoning, they helped set a new tone in astronomical research.

Shapley's three-dimensional map of the world of globular clusters showed that they were arranged in a vast spherical volume of space, as if they themselves were members of a sort of super globular. The center of this sphere was nowhere near the sun, but instead lay tens of thousands of light-years away in the direction of the constellation Sagittarius.

Shapley now made an intuitive leap and asserted that the center of the world of the globular clusters was the center of our stellar system, the Milky Way, as well. Copernicus had found that Earth was not the center of the universe; Shapley found the sun wasn't either. We live, he saw, in our galaxy's suburbs.

Shapley was right about our place in the Milky Way, but wrong about its size. He estimated the diameter of our galaxy at 250,000 light-years, with the sun situated 50,000 light-years from its center. The correct values are more nearly 100,000 light-years diameter, with the sun 30,000 light-years, or about three-fifths, out toward one edge. Shapley had failed to take into account that interstellar dust and gas dim our view across the galaxy, especially close to the galactic plane, which was just where most of his globular cluster studies led. Measuring a giant star in a remote globular, he attributed its dimness to sheer distance, when in fact it was caused in part by absorption of light by interstellar dust and gas. As a result he tended to put the clusters, and consequently the center of the galaxy, too far away.

The mistake was not crippling so far as understanding our own galaxy was concerned, but it got Shapley into difficulty when discussion turned to the question of whether other galaxies exist. Those who thought, quite correctly, that the Magellanic Clouds were satellites of the Milky Way, found Shapley's Milky Way so large it engulfed the Clouds. Similarly, it was hard to see how the spiral "nebulae" could be separate galaxies, unless ours was somehow a bloated monarch and all the others dwarfs. Shapley took an unyielding stand in defense of his picture of the Milky Way, inflated figures and all. He denied that the spiral nebulae were other galaxies. Since the spirals were the best candidates for the role of galaxies, he seemed to be implying that our Milky Way was all there was to the universe.

From Lick Observatory, up the California coast near San Jose, an astronomer named Heber Curtis began sniping at Shapley's work. This was an early volley in a long war. Lick Observatory was associated with the University of California, Mt. Wilson with the Carnegie

Institution and later Cal Tech, and the two were to quarrel for years. Curtis judged Shapley's model of the Milky Way ludicrously outsized. He maintained the spiral nebulae were other galaxies like ours.

A debate between the two astronomers was arranged by the National Academy of Sciences in Washington, D.C. Shapley and Curtis rode the Southern Pacific east together, restricting themselves to nonscientific small talk so their debate would be fresh. It took place April 26, 1920, with Albert Einstein in the audience.

Curtis focused on Shapley's weakest point, his contention that the spiral nebulae were cosmically local. Shapley took the bait and brought forth two spurious arguments. He dredged up the 1885 Andromeda Nebula supernova, arguing that it proved the spiral could not be a galaxy because that would mean a single exploding star could equal the light of hundreds of millions of ordinary stars, which struck him as an obvious absurdity. But supernovae do indeed shine that brightly. Shapley then cited observations by his friend and colleague, Adrian van Maanen, indicating that the Andromeda spiral was drifting across the sky at a clip that, assuming it were at extragalactic distance, would require it to be moving faster than light, something Einstein had shown only a few years earlier to be impossible. But Van Maanen's data were incorrect.

In retrospect, Shapley's arguments, while flawed, came closer to the truth than Curtis's. Shapley was correct that the sun is located well out from the center of our galaxy, while Curtis clung to an earlier notion that we sit in the center of things. Shapley's understanding of globular clusters was the more sophisticated, though he erred in his values for the intrinsic brightness of their stars. And while Shapley's model of the Milky Way was too large, Curtis's, by an even wider margin, was too small. Both men agreed that absorption of starlight by interstellar dust and gas was negligible; on that point both were wrong.

In any event, the impression was widespread that Curtis had defeated Shapley, and Shapley found himself widely judged the loser in what ranked, after the Scopes trial, as the most widely publicized American scientific debate of the time. He never forgot it. Decades later he was still fuming at Curtis for, he felt, drawing him off the subject, and at Van Maanen for botching the Andromeda Nebula observations. In his memoirs he wrote, "They wonder why Shapley made this blunder. The reason he made it was that Van Maanen was his friend and he believed in friends!"

Soon after the debate Shapley quit Mt. Wilson. He accepted the directorship of Harvard College Observatory (Pickering had died in 1919). Friends urged him to reconsider; compared to Mt. Wilson, where the heavens were explored nightly with the best telescopes available, Harvard was a counting house. But Shapley went off to Cambridge. He became a distinguished figure, president of both the American Academy of Arts and Sciences and the American Association for the Advancement of Science, winner of a drawerful of medals, author of a half-dozen popular books on astronomy, a radio personality. He dabbled in politics, helped found UNESCO, and was attacked by Senator Joseph McCarthy. When the spiral "nebulae" proved to be galaxies after all, the observatory graduated from cataloguing stars to cataloguing galaxies, by the thousands, with the help of new generations of women assistants for whose toil Shapley coined the term "girl hours."

While Shapley kept busy, the real breakthroughs were

being made at Mt. Wilson. If this pained him he never said so, though he conceded, "sometimes I have thought that anything I did after that [Mt. Wilson] was anticlimactic." Shapley read in the technical journals that back on the drafty peak of Mt. Wilson, Edwin Hubble—Hubble of all people, cold and overbearing Hubble—was discovering the universe.

Shapley disliked Hubble. The man was impolite; when an article of Shapley's was sent him for checking, he scrawled "Of No Consequence" across the manuscript and returned it, and the comment was inadvertently set in type under Shapley's byline. He was ungenerous: "The work that Hubble did on galaxies was very largely using my methods. . . ." Shapley wrote. "He never acknowledged my priority, but there are people like that." He seldom asked advice. He spoke with an Oxford accent which especially infuriated Shapley, who, like Hubble, came from Missouri. Shapley strongly suspected that if Hubble were awakened in the middle of the night he would again talk like a Missourian.

A tall man with a strong jaw, thin mouth and chilly gaze, Hubble did little to court the affection of his colleagues. A handful of friends found him warm and even charming, but to others he appeared arrogant and prepossessing. A photograph taken of him dressed for trout fishing, flyrod in hand, staring into the camera, makes one feel for the trout. Most would admit, but few volunteer, that he was one of the greatest astronomers who ever lived.

Born in 1889 in Marshfield, Missouri, the fifth of seven children raised by a disciplinarian father, Hubble read Jules Verne and learned the constellations as a boy. He went to the University of Chicago on a scholarship and emerged as a quick student, a track and basketball letterman, and a skillful enough boxer to fight an exhibition with Georges Carpentier, the world light-heavyweight champion. Promoters looking for a "white hope" to fight Jack Johnson, the black heavyweight, asked Hubble to turn professional, but he chose instead to study law at Queens College, Oxford, as a Rhodes scholar. In 1913 he joined the Kentucky bar and practiced for a few months in Louisville, but the law bored him. "Astronomy mattered," he said.

Hubble re-enrolled at the University of Chicago, which operated Yerkes Observatory in Williams Bay, Wisconsin, near the Illinois border, home of the world's largest refracting telescope.* Built in 1897 with a donation from Charles Tyson Yerkes, the ex-convict Chicago streetcar baron, this instrument, with its 40-inch lens, marked the climax of the age when astronomical telescopes resembled oversized spyglasses. Lilliputian astronomers toiled beneath its towering pier and 36-foot riveted steel tube; to reach the eyepiece they climbed bleacherlike wooden scaffolds that, with a little wrestling, could be moved around the observatory floor as the telescope tracked stars. Hubble, despite a few interruptions—he was stabbed in the back by a would-be thief in the woods one summer but not badly injured, and once dove into Williams Bay to save a woman from drowning—got his Ph.D. in astronomy at Yerkes. In World War I, he enlisted in the infantry, was wounded in France

---

* The cardinal virtue of an astronomical telescope is to gather light. Refracting telescopes, from Galileo's through to the Yerkes refractor, did this by means of a lens. But the weight and thickness of the glass renders impractical refractors of more than about 40 inches diameter. Therefore the major telescopes of the twentieth century have been reflectors, using a curved mirror to gather the light and bring it to a focus. The mirror can be supported from the back for structural integrity, and less light is lost than with a refractor. Several reflecting telescopes with mirrors 200 inches or greater in diameter have been built.

by shell fragments in the right arm, and finally arrived at Mt. Wilson in 1919, lured by the new 100-inch telescope.

From the outset Hubble's preoccupation was the spiral nebulae. He wrote his graduate thesis on the subject, arguing that astronomers ought to proceed on the assumption that the spirals were galaxies, since that line of inquiry, if justified, would lead to great things.*

Evidence that he was on the right track had come in July 1917, when Mt. Wilson's expert optician George Ritchey noticed a point of light on an old photograph of a spiral nebula in Cygnus and correctly surmised it was a nova. Ritchey and Curtis then searched photos of the Andromeda Nebula in the files of both Mt. Wilson and Lick observatories and found two previously unnoticed novae in plates Ritchey himself had made. Luckily both appeared in sequences of photos made on more than one night, so the light curve of each exploding star—the way it brightened, then dimmed, with passing time—could be reconstructed and compared with the behavior of novae that had been observed in our galaxy. There was clearly a resemblance, and the relative dimness of the newly discovered exploding stars constituted the best evidence yet that the Andromeda spiral must be very far away.

Hubble first studied nebulae he thought were probably local. Some of these were objects he had known since boyhood, among them the Pleiades and the blossomlike Orion nebula. Others were dimmer. Using Mt. Wilson

* Hubble may have borrowed this line of thought from the English astronomer Arthur Stanley Eddington, who wrote in a book published two years earlier, "If the spiral nebulae are within the stellar system, we have no notion of what their nature may be. That hypothesis leads us to a full stop. . . . If, however, it is assumed that these nebulae are external to the stellar system, that they are in fact systems coequal with our own, we have at least an hypothesis which can be followed up, and may throw some light on the problems that have been before us."

telescopes from the little 10-inch Cooke camera up to the 60-inch reflector, after five years he had sorted the nearby, or "galactic," nebulae into distinct categories.

In a 37-page paper submitted May 1922 to Hale's *Astrophysical Journal,* Hubble first paid homage to nebulae observers as far back as Herschel, then set forth his own findings. "The nebulosity seems to consist of clouds of matter, molecules, dust, or perhaps larger particles, not hot enough to be self-luminous, but visible because of light excited by or reflected from involved or neighboring stars," he wrote. Some nebulae shine by the reflected light of nearby stars. Others, if the star involved is hot enough, are bombarded by starlight so intense that their molecules emit light of their own, like those in a fluorescent lamp. Hubble showed that when ordinary stars were involved the spectrum of the nebula looked like that of a star; we were simply seeing reflected starlight. If the stars were bright enough, the nebula were "excited" (as the spectroscopists say) into producing spectra of their own. The result was an emission spectrum, so called because the nebula itself emits, rather than just reflects, light. This explained why the Pleiades displays a starlike spectrum, while the Orion nebula, which looks much the same to the eye, has an emission spectrum.

Hubble showed that emission nebulae like that in Orion were always associated with very hot stars. Since the stars were certainly within our galaxy, these nebulae must be too. Hubble therefore listed them as "galactic." The spirals, the globular clusters, and some elliptical and irregular nebulae he classified "non-galactic."

Though he doubtless felt all but certain that his "non-galactic" nebulae were galaxies like ours, and that in investigating them he was therefore embarking upon an exploration of unprecedented expanse, Hubble avoided saying so, seeing no reason to become embroiled in the Shapley-Curtis controversy. "There appears to be a fundamental distinction between galactic and non-galactic nebulae," he wrote, but quickly added with a bow to Shapley's side, "This does not mean that the latter class must be considered as 'outside' our galaxy." Years later, when he was being acclaimed as the man who unveiled the galaxies and opened our eyes to the universe, he still stuck to his old terminology. "The term *nebulae* offers the values of tradition," he wrote, "the term *galaxies,* the glamour of romance."

To resolve one of the spirals into stars would establish without much doubt that it was a galaxy, but this task was too much for the 60-inch telescope. A few of Ritchey's superb photographs did seem to reveal a scattering of individual stars in some spirals, but Ritchey himself cautioned that this evidence was tentative at best; the spirals could be nearby swirls of gas, as Laplace had envisioned, with a few new stars embedded in them. When the 100-inch came into operation, Hubble made photos that came tantalizingly close to resolution. Examining the spiral nebulae under a magnifying glass, Hubble sometimes felt he was seeing only a smooth wash of light. Other times, depending, he said, "on what I had for breakfast," he felt certain that the millions of tiny grains in the soupy imprint of the spirals were not a photographic effect or his imagination, but stars, that he was staring at a vast continent of stars. Ultimately the 100-inch telescope would solve the riddle of the spirals, but for the present the question could not be answered.

One prophetic if inconclusive inquiry came from an astronomer named E. Oepik, who in 1922 published a paper arguing that the Andromeda spiral was 1.5 million light-years away. This was, for its day, remarkably close (the modern estimate is between two and two and a half million light-years), closer, in fact, than Hubble got for another twenty years. But it won little support.

Hubble, denying himself any such "unverified speculation," as he called it, stuck to solid research, following several avenues at once. Like Shapley, he searched for Cepheid variable stars as a way of estimating cosmic distances. He found them in several non-galactic nebulae.

For a while Hubble concentrated his search on an odd clutch of stars labeled NGC 6822. This assembly had first been noted in 1886 by Edward E. Bernard, a self-educated American astronomer, and was studied more closely by Charles Perrine, a British veteran of Lick who settled in Argentina to view the southern skies. Located near the rich star clouds of Sagittarius, it might have been mistaken for part of the Milky Way, but Perrine, an astute observer, realized it was a distinct entity, inhabited by a variety of indigenous stars. If this was so, Hubble realized, it might be a galaxy close enough for ready examination. Over a period of two years he photographed NGC 6822 and found 50 Cepheids. Their brightness, analyzed largely with Shapley's methods, put them several hundred thousand light-years away. Here clearly was an object independent of our galaxy.

NGC 6822 seemed so staggeringly distant that Hubble was relieved to find identifiable stars there; he had not been certain other galaxies would even obey the same physical laws as ours. Finding the Cepheids reassured Hubble that observing deep into the cosmos, while difficult, was not going to be impossible. "The principle of the uniformity of nature . . ." he wrote, "seems to rule undisturbed in this remote region of space."

Having tested his skills in the cosmic archipelago, Hubble steered for deep water. For several years he had investigated the two largest spiral nebulae visible in Earth's skies. These hundreds of observations now generated two long papers, each a legion of research, bristling with data— one on the nebula M33 in Triangulum, in February 1926, the other on the Andromeda Nebula, in December 1928.

M33 is a lovely spiral, oriented in space so we view it almost flat-on, pretty as an ocean sunfish. Using a new, more sensitive photo emulsion, Hubble photographed it repeatedly with the 100-inch telescope on the clearest nights and finally succeeded in resolving it indisputably into stars. He then identified 35 of the stars as Cepheid variables (only three had been located before, and those tentatively) and from them obtained a distance estimate that put M33 even farther away than NGC 6822, establishing it as a spiral galaxy, separate from and outside the Milky Way.

Hubble's Andromeda Nebula paper virtually broke the back of the subject. Previous astronomers had studied dozens of photographs of this entrancing spiral. Hubble scrutinized 350, of which he took more than 200 at the telescope himself. Two novae had been found there, by Ritchey; Hubble discovered 63, and reached the remarkable conclusion that the Andromeda is so populous a galaxy that 30 stars in it explode every year. He estimated its size and mass, and guessed correctly, though perhaps for the wrong reasons, that it might eventually prove to be still larger.

With the publication of these papers the peoples of our planet discovered what, presumably, most cognizant beings evolving beneath clear skies learn sooner or later—that the cosmos is subdivided into galaxies.

The discovery, a long time in coming, was followed immediately by that of the expansion of the universe.

Hubble had evidence of expansion on his desk while still proofreading his Andromeda Nebula paper. In the course of establishing the approximate distance, size and brightness of dozens of galaxies, he had also measured the velocity of each relative to us. His original purpose was hardly more ambitious than to find out how rapidly the sun was moving as it took part in the slow spin of our galaxy; like passengers on a carousel, we ought to be able to determine how fast we are going by looking at fixed objects in the distance. Hubble assumed other galaxies would provide such points of reference, whether they were fixed or drifted at random in space. The sun's motion in our galaxy could be deduced from the apparent motion of the other galaxies, some of which presumably would appear to be approaching if we were headed their way, others apparently to be receding. Hubble did find traces of this effect, but overshadowing it he discovered something unexpected, monolithic and strange. Only a few local galaxies showed the sort of random drift anticipated. All the others appeared to be moving away. They were receding at remarkably high speeds, and, most startling, the farther away they were, the faster they were receding.

There were two uncomplicated ways to explain this state of affairs. One was to say that our galaxy sits at the center of the cosmos and all the others are, for some reason, rushing away from us, and they run faster the farther they get. This seemed unlikely. The other explanation was that the universe is expanding. In any uniformly expanding entity—a spotted balloon, a loaf of rising raisin bread, or a cosmos of galaxies—the distance between two given points will increase with time, and the greater the distance to start with, the faster it will increase. The British astronomer Eddington illustrated this effect by asking his students to envision what would happen if the lecture hall they were sitting in (the universe) were to double in size, carrying their desks and chairs (the galaxies) with it. Each would notice that nearby students had to move their desks only a few feet in order to double the distance, while those across the room had to move much farther; in any given period of time, therefore, the farther one looked across the hall, the faster one saw desks and chairs receding.

Hubble estimated the velocities of galaxies by breaking down their light with a spectrometer and measuring the degree to which lines in their spectra were offset. If a galaxy were approaching, its light would be crowded a bit toward the high-frequency end of the spectrum, the blue end. If it were receding, the light waves would be stretched out, shifting the lines toward the lower-frequency red end. In all but the most nearby galaxies Hubble observed, the spectral lines were displaced toward the red. The rule could be stated: The more distant a galaxy, the greater the redshift of its light.

The historical scaffolding under Hubble's discovery dated back to the early nineteenth century, when Christian Doppler, a physicist teaching in Vienna, found that the wavelength of lines in the spectrum of a light source ought to shift if the source were moving rapidly toward or away from the observer, just as the pitch of an automobile horn sounds higher—shorter wavelength—if the car is approaching and lower—longer wavelength—if it is speeding away. The spectrum is said to be "Doppler shifted" by velocity. Spectral lines could be measured with considerable accuracy, so this meant the velocities of remote, bright objects, namely stars, could be too. James Keeler at Lick

Observatory quickly succeeded in finding a Doppler shift in the spectrum of the bright star Arcturus: it had a blue shift, indicating it and the sun were growing closer in the course of their common sweep in our galaxy. Thereafter Doppler shifts were widely employed in finding the velocities of stars.

The first sustained attempt to measure Doppler shifts in the light of galaxies came about prior to Hubble as an inspiration of the eccentric astronomer Percival Lowell. Lowell came from a prominent Boston family—his great-great-grandfather John was a member of the Continental Congress, his brother Abbott president of Harvard, his sister Amy the poet and biographer of Keats—and he combined a passion for astronomy with the money to do something about it. On a pine-shrouded peak outside Flagstaff, Arizona, he built Lowell Observatory, and there embarked on a scientific career informed by motives as much poetic as scientific. Lowell was convinced that the canals of Mars, since found to be an optical illusion, were waterways constructed by a Martian civilization perishing of thirst, and he insisted, just when most professional astronomers were discarding the idea, that the spiral nebulae were solar systems in the making. This devotion of his to an outmoded theory of the nebulae led circuitously to important work presaging Hubble's.

Lowell hired, perhaps intentionally as a counterbalance to his own flamboyance, an observer named Vesto Slipher. He later made him observatory director. Patient and fastidious as Lowell was impulsive, Slipher never cut corners and never published a result until he was sure it was correct. He wore a suit and kept his tie knotted even when alone at the telescope all night. Lowell encouraged him to make spectra of the spirals and look for Doppler shifts, expecting the results would support the old Laplace theory. Slipher himself felt this was unlikely and thought the spirals were going to turn out to be galaxies, but he dutifully set to work taking spectra of them and measuring Doppler shifts, repeating difficult observations many times to insure accuracy. He had the field virtually to himself for more than a decade (Lowell was involved in it for the wrong reasons, and nobody else wanted to get into it much at all) and by 1925 he had obtained Doppler shifts for 45 spirals.

Slipher was not surprised that his results failed to support Lowell's hypothesis. But he was surprised to find the nebulae not moving randomly in space but nearly all receding rapidly. The redshift of an average galaxy on his list showed it receding nearly 400 miles per second. Some were moving 1,000 miles per second relative to us.

A few astronomers who saw these data, notably Carl Wirtz in Germany, perceived something of a correlation between the distances of the spiral nebulae and their velocity of recession, but the idea remained tentative in the absence of more information on distances. The man who knew those distances was Hubble. In 1927 the sun's velocity of rotation in our own galaxy was finally measured fairly accurately, and it became possible for Hubble to see clearly that the galaxies behaved according to a grand and startling design—the farther away, the more redshift.

Characteristically, Hubble handled his discovery with long tongs. Word of what one astronomer later called "the most astounding fact of the twentieth century" came in a modest little paper that contained no mention of the expansion of the universe or, for that matter, of the universe. Titled "A Relation Between Distance and Radial Velocity Among Extra-Galactic Nebulae," it approached the subject via its original and most mundane application, as a complication in the problem of fixing more accurately the sun's velocity in our galaxy. "New data to be expected in the near future may modify the significance of the present investigation or, if confirmatory, will lead to a solution having many times the weight." That was as evocative as Hubble got, in announcing that the universe, long assumed to reside in stately lassitude, was instead expanding from genesis toward destiny.

For all his caution, Hubble saw from the outset that the velocity-distance relation, if it held up, afforded him a powerful new way to chart deep space. The Cepheid variable stars had yielded distances to several galaxies. By studying bright giant stars in those galaxies (as Shapley had done with the globular clusters), Hubble was able to measure the distances of systems where the Cepheids were too faint to be seen. Beyond that, he had obtained some distances by using the total brightness of whole galaxies as his standard, but this method was rough. The velocity-distance relationship promised a yardstick of greater length. If it held up, distances to galaxies almost too remote to be photographed could be inferred from their redshift. The velocity of their recession would itself inform astronomers of their distance.

Increasingly from 1928 on, Hubble entrusted the night-by-night task of exploring these uncharted reaches to Milton Humason, an unusual man. Humason showed up at Mt. Wilson while traveling around California in search of work. He was hired first as a mule packer, then as busboy in the mountaintop dining room and as observatory janitor. He asked questions constantly. Shapley and some of the other astronomers, impressed by his curiosity and intelligence, had him promoted to night assistant on one of the smaller telescopes. Humason did the job with energy and wit, and managed to wangle a few routine observing assignments of his own on the little 6-inch telescope, then the 10-inch. Hubble finally tried him out on the 100-inch and was pleased with the results. Because Humason had only an eighth-grade education, Hale long denied him another promotion. But he finally capitulated and made him a full staff member. Easygoing and unaffected, Humason played poker with the night assistants and joked with Einstein in the same unassuming manner. He became liaison between the olympian Hubble and the other astronomers, and served as Hubble's assistant at the telescope, shouldering the bulk of the observing while Hubble smoked his pipe and pored over the results. First he installed a specially designed "fast" lens at the eyepiece end of the 100-inch.* After testing it on previously observed galaxies he moved out into deeper space, where entire clusters of galaxies were, by Hubble's plan, to be assayed like grains in a prospector's pan. The telescope was trained on the Pegasus cluster, the Perseus, Coma, and Bootes clusters; and a science that a few years before had not been certain whether the cosmos ended with the Milky Way now charted territory hundreds of millions of light-years out.

On clear, moonless nights, Humason perched on a little steel platform 50 feet above the observatory floor, the black skeleton of the telescope tube silhouetted against the dome slit, stars and sky. A night assistant, holding Humason's old job, sat at the console, his face illuminated dimly by red dark-vision lamps. They worked to a continuo of mountain winds and the heavy ticking of a weight-driven brass clock that powered the telescope. Intermittent chimes signaled

* Photographers will appreciate that it was F 0.6.

that Humason, up in the dark, had pressed a button altering the drive rate to keep his guide star in place. If the correction mechanism got balky, as it often did, Humason would hold the image in place by leaning on the massive telescope, sometimes climbing aboard it; he would perform whatever acrobatics were required to hold the field steady, while the photographic emulsion steeped in the ancient light of galaxies.

These nights tested Humason physically as well as intellectually. An observatory dome cannot be sealed from the elements because glass windows would steal precious light; it cannot be heated because the heated air would rise, setting up air currents, and the slightest air turbulence scrambles the image in a large telescope. In winter the observer huddles at the eyepiece, tears streaming down his cheeks, breathing deeply to ward off attacks of shivering, staring at a guide star, which he cannot ignore for a moment without risking his time-exposure photograph or spectrum. As the telescope tracks across the sky the eyepiece can roll to diabolical orientations, prompting recollection of a Gilbert and Sullivan-style tune Shapley and his Harvard students composed to describe the ideal astronomer:

> His knee should bend and his neck should curl,
> His back should twist and his face should scowl,
> One eye should squint and the other protrude,
> And this should be his customary attitude.

The respected Rudolph Minkowski once endured five successive nights of this punishment exposing a single plate in an effort to record a very faint galaxy, then at dawn after the fifth night, exhausted, went down to the observatory darkroom to develop it, where by mistake he immersed it not in developer but in fixer, destroying the image. He stayed in astronomy anyway.

By the time Humason reached the Bootes and Ursa Major clusters, he was dealing with galaxies too faint to be seen by eye even through the 100-inch telescope. His technique was to locate them in wide-field time-exposure photographs, then aim the big telescope by setting it on the proper celestial coordinates, hoping the galaxies were there in the black field. The setting circles employed for this were enshrined high up in the telescope mount; the night assistant read them from the console through two small telescopes of his own. Humason had to wait each time until the plate was developed before knowing if he had caught anything. In this fashion, he found galaxies receding at 26,000 miles per second, a seventh the velocity of light.

New, faster photo emulsions made it possible to record still more distant galaxies in the same exposure time, and Humason's plates showed conventions of galaxies thick as protozoa. An informal goal of Hubble's was to make a photograph that recorded as many galaxies as foreground stars. This was achieved on March 8, 1934. The print looked like beach sand scattered on a sheet of paper. About half the dots were stars in our galaxy; the others were galaxies, each home to billions of stars of its own. Hubble called it "a spectacular description of the penetrating power of the telescope." If he felt anything more, he did not say so.

2

WHY IS
THE SKY DARK
AT NIGHT?

It requires a very unusual mind to
undertake the analysis of the obvious.

—ALFRED NORTH WHITEHEAD

WILHELM OLBERS WAS A GERMAN PHYSICIAN WHO ENJOYED sweeping the night sky with a small telescope, looking for comets and asteroids. He complemented this restful way of passing time with a tenacious grasp of orbital mechanics. He calculated the first accurate orbit of a comet, discovering that comets originate in an exurban dump of our solar system out beyond the orbits of the planets. He also learned that the asteroids, lumps of rock ranging from the size of a baseball to chunks larger than cities, come from a belt of debris lying between the orbits of Mars and Jupiter. These discoveries made Olbers famous, but he stuck to medicine, exploring the stars as a diversion.

In 1826, at age sixty-eight, Olbers wrote a little paper posing one of the unsettling questions of scientific history. Why, he asked, is the sky dark at night?

Through the eighteenth and nineteenth centuries the universe was widely presumed to be infinite. This view deferred the dilemma of how a three-dimensional universe could be finite (if it came to an end, what lay beyond?) and had the endorsement of Sir Isaac Newton. If the number of stars was finite, Newton wrote in 1692, their combined gravity would pull them together; they would "fall down into the middle of the whole space, and there compose one great spherical mass." The alternative to that uninspiring state Newton saw to be an infinite universe peppered with infinite stars. "If the matter was evenly disposed throughout an infinite space it could never convene into one mass," he wrote. ". . . Some of it would convene into one mass and some into another, so as to make an infinite number of great masses, scattered at great distances from one another throughout all that infinite space." And indeed Newton's "great masses" sound something like galaxies.

In his deliberations, Newton relied upon euclidean geometry, the one workable geometry known; twenty centuries of surveying, navigation, building and teaching testified to Euclid's genius, and his system was widely regarded as little less than revealed truth. Euclidean geometry implies the existence of infinite space—parallel lines must extend to infinity before they meet, for example—and so, when Newton conceived of the cosmos in euclidean terms, he naturally was inclined to think of it as infinite.

That satisfied almost everyone until Olbers. Why is the sky dark at night?

If there really are an infinite number of stars, we ought to see stars crammed into every niche in the sky. Imagine that a small telescope shows two stars in a given field of view. A more powerful telescope should reveal stars filling the space between them, and a still larger telescope should find stars filling the spaces between *them,* and so on until all we see are stars and no sky whatever. In other words, if Newton's model of the cosmos were correct, every line of sight in the sky, no matter how narrow, must eventually encounter a star. Prior to Olbers it had been assumed that most of the infinite stars were invisible simply because they were too far away for us to see them. (The apparent brightness of a star diminishes by the square of its distance, so that if one star is 100 light-years from us and another identical star is 200 light-years away, the second star will appear to us only one-quarter as bright as the first.) Olbers examined the matter more critically, and found that contrary to what had been assumed before, the cumulative brightness of an infinite number of stars should overwhelm the dimming effect of their distance. Olbers realized that if we do live in a static, homogeneous, infinite universe, as Newton had maintained, the night sky ought to be a blinding sheet of light, bright as the surface of the sun. Yet it is dark.

This is known as Olbers' paradox. Actually much the same point had been made before, in 1744 by a Swiss astronomer, J. P. L. de Cheseaux. Olbers had a copy of Cheseaux's book in his library but apparently never read it, and most scholars credit him with having arrived at the idea on his own. Edmund Halley, known for the bright comet named after him, stumbled on a form of the paradox even earlier but seems not to have recognized its significance.

Olbers considered and rejected two ways of escape. One was to assert that stars far from the sun get intrinsically dimmer, but this only begged the question. Another was to say the universe was infinite but the number of stars was not. Perhaps the cosmos was composed of the Milky Way and empty space, nothing more. Olbers rejected that possibility because, according to Newton, the Milky Way would collapse if it were true. (Subsequent analysis indicated Newton was wrong on this point, but since we now know the Milky Way is not the only galaxy—far from it—the paradox endures.) Olbers finally decided that interstellar matter probably absorbed the light from distant stars. This was astute in a sense—Shapley might have profited by taking interstellar matter more seriously—but turned out to be insufficient. Given infinite time, starlight will heat up interstellar dust and gas until it glows and the universe will be bright as the sun after all. The laws of thermodynamics, derived after Olbers' death, made that clear and renewed the paradox.

Olbers' paradox was not resolved until Hubble discovered the expansion of the universe. The sky, it seems, is dark because the universe is expanding. To answer Newton, the universe does not collapse because it is expanding. To answer Olbers, it is not dazzlingly bright because it has something else to do with its energy—namely, expand. The remote galaxies dump so much of their light into the redshift, so much of their energy is thinned out over increasing space and time, that the sky looks dark.

In recent years some theoretical physicists have implicated the expansion of the universe in matters as diverse as the direction time flows and why a teapot full of cold water must be heated to make it boil; the laws of thermodynamics, in their view, result from cosmic expansion. These hypotheses, if valid, enrich the insight we identify with Olbers, that the universe is close to home as well as distant, and that so familiar a sight as the dark of night results from so grand a thing as the motion of the galaxies.

3

THE UNIVERSE OF THE MIND: COSMOLOGY

It is obvious that we must regard the universe as extending infinitely, forever, in every direction; or that we must regard it as not so extending. Both possibilities go beyond us.

—EDITORIAL IN *Scientific American*
MARCH 13, 1921

STUDENTS OF THE SKY FROM LUCRETIUS TO NEWTON WHO believed the cosmos to be infinite appear to have held that opinion less because they were enthusiastic about it than because the alternative seemed worse. A finite, three-dimensional universe would necessarily come to an end somewhere, an unimaginable situation. As a conversant of the Chinese philosopher Wang Ch'ung asked 2,000 years ago, in much the same words still heard today, "If Heaven has a boundary, what things could be outside it?" Lucretius regarded the notion of a finite universe as ridiculous. "Let us assume for the moment that the universe is limited," he wrote. "If a man advances so that he is at the very edge of the extreme boundary and hurls a swift spear, do you prefer that this spear, hurled with great force, go whither it was sent and fly far, or do you think that something can stop it and stand in its way?"

The strongest arguments against a finite cosmos have been the sort of reductions to absurdity exemplified in Lucretius. But the idea that the universe is infinite proves to be just as paradoxical. If we suggest that there are an infinite number of stars and galaxies, we must deal with a finding by Carl Neumann and Hugo Seeliger that gravity as we know it could not operate in such a universe; gravity from the infinite remote masses, the two mathematicians learned, would overwhelm local gravity at every point in the cosmos. In addition, the concept of an infinite quantity of any *thing* raises intuitive problems. Alexander the Great purportedly wept upon being told infinite worlds exist because, he complained, "there is such a vast multitude of them" and he had conquered only one. He might have consoled himself with the thought that a cosmos of infinite worlds might contain infinite numbers of Alexanders, busy conquering infinite Earths. Still, all their triumphs would amount to only an infinitesimal fraction of the whole, so perhaps our local Alexander sobbed justifiably. There is something unreasonable about the whole line of thought. If, instead, we propose that only space is infinite—that galaxies and stars cease populating the universe at some distance and that beyond ranges empty space—we find it difficult to explain what is meant by "space" in reaches infinitely beyond matter. Space in the absence of objects is, if you think about it, indistinguishable from no space at all, a conclusion reached by Newton, Aristotle and George Berkeley. Infinite and finite universes trouble the mind equally.

I will now say that this ancient problem may have been solved in our century. One should be suspicious of claims that tough old questions recently have been answered. Still, some new things do appear under the sun, and the concept of a finite but unbounded universe was one. In a series of developments beginning in the nineteenth century and reaching fruition with the theory of general relativity in the twentieth, this idea released cosmology, the discipline concerned with the physical nature of the universe as a whole, from the finite/infinite dilemma. Two broad currents of thought led to it: the scientific positivism of Ernst Mach and the non-euclidean geometries of Karl Gauss, János Bolyai and Nikolai Lobachevski. They converged in Einstein.

First Mach. He was born in Turas, Moravia, in 1838. A poor student, he struggled a few years at an Austrian

Benedictine academy, withdrew, served as a cabinet-maker's apprentice, returned to school at a Moravian gymnasium where he hated the classes but endured and finally gained admittance to college. Mach "looked into the world with the curious eyes of a child," Einstein wrote; a sense of the relativity of things seemed wired into his perceptions. Looking down from a bridge at a flowing stream, he became giddy with the feeling that the bridge was moving. A steamboat trip on the Elbe made him dizzy because he felt the land, not the boat, was gliding along. He frowned at the shabby dress of a man he saw on a train, then realized it was his own image in a mirror.

These uncensored perceptions, discomfiting in adolescence, served Mach well as an adult. He did distinguished work as a psychologist of human perception, a physicist in the propagation of sound (the speed of aircraft relative to that of sound is measured in Mach numbers), and a philosopher who insisted that scientists stick to data and shun assertions about absolute reality, or lack of it, in their theories. Scientific "laws," Mach argued, result from the common experience of human beings and deserve promotion to no higher office. "Physics is not the whole universe. . . ." he wrote. "Our physical concepts, however close they come to the facts, must not be regarded as a complete and final expression of these facts." The "self-evident" postulates of Euclid and the "absolute" space of Newton, Mach felt, should be treated as no more than useful tools, to be discarded without remorse if tools better fitting the data were found.

In this, Mach was influenced by the eighteenth-century empiricist Berkeley, who felt the essential quality of existence of a thing was our perceiving it, and so had little use for absolute space. In Berkeley's *Principles of Human*

Knowledge, Mach read these words: "Absolute space, distinct from that which is perceived of sense and related bodies . . . cannot exist without the mind. . . . We cannot even frame an idea of *pure space* exclusive of all body." Imagine that only a single sphere exists alone in the universe, Berkeley wrote. We cannot envision that sphere revolving, because we have no reference points, nothing beyond the sphere to reveal that it *is* revolving. Imagine two globes in an otherwise empty cosmos. Can they whirl round one another, like the rubber tips of a spinning baton? Again the idea is meaningless without some exterior point of reference. Newton's absolute space lends to our mind's eye little help. "But suppose that the sky of the fixed stars is created," Berkeley wrote. "Suddenly from the conception of the approach of the globes to different parts of that sky, the motion will be conceived."

At the time, Berkeley's argument was lost in Newton's shadow, but Mach, reading it a century and a half later, was fascinated. He thought about a famous argument of Newton's concerning centripetal force. Hang a bucket of water by a rope and spin it; the water climbs up the inside of the bucket while depressing somewhat in the center. This behavior, according to Newton, results from the water's motion relative to absolute space. The more Mach thought about this position of Newton's the less tenable it appeared. What reality could be accorded to invisible, intangible space such that it could dictate terms to the water? Carrying Berkeley a step further, Mach reasoned that if the bucket could be said to be spinning only relative to some exterior point of reference, then the force that made water in the bucket climb its walls required those reference points too. "I have no objection to calling [it] rotation," Mach wrote, "so long as it be remembered that nothing is meant except relative rotation with respect to the fixed stars." He had no objection to talk of "centripetal force," so long as it was kept in mind that without the distant stars, no such force could be said to exist.

Mach's seemingly abstract point contained the seeds of a profound picture of a cosmos where objects, instead of pushing and pulling one another by force, engage in a sort of silent collaboration that reads out to us as "laws" of nature. Inertia, the fundamental quality of matter from the standpoint of traditional science, was for Mach not inherent in each object, but a reflection of its relationship with all the other objects in the universe. How hard you must push a boulder to get it moving, the force a batter needs to hit a home run, whether a coin stops rolling before falling down a grate, are matters bartered among the stars. Though Mach did not explain, nor has anyone since, how his cosmic interaction actually worked (it has never been mathematically formulated), Einstein was inspired by it. He called it "Mach's principle" and attempted to incorporate it into his then embryonic theory of general relativity.

Einstein visited Mach one day in autumn 1913 when Mach was seventy-five years old, semiparalyzed, and sufficiently hardened in his views that he denied the existence of atoms. Atomic theory was revolutionizing physics at the time, but Mach regarded the "atom" as just another concoction of the mind, useful as a conceptual tool but lacking independent reality. Three years earlier, Mach had written bitterly to Max Planck, "If belief in the reality of atoms is so important to you, I cut myself off from the physicist's mode of thinking. . . . I decline with thanks the communion of the faithful. I prefer freedom of thought." Cut off he was. He received few visitors and when they appeared in his small apartment in a Vienna suburb he shouted, "Please speak loudly to me. In addition to my other unpleasant characteristics I am also almost stone deaf." Einstein sat and shouted pleasantly about atoms.

There is no record whether the two talked about Mach's views of inertia and space, but the matter had captivated Einstein since his student days at the Federal Institute of Technology in Zurich. Just before visiting Mach, Einstein hinted in a speech that he was at work on a theory of gravitation that would be as far removed from Newton, he said, as radio was from Ben Franklin's kite and key.

The non-euclidean geometries, another antecedent of general relativity theory, originated in the failure of a hundred generations of mathematicians to derive Euclid's axiom of parallel lines. The axiom implies, in a roundabout way, that if we draw a line AB on a plane and a point P not on that line, one line and one line only may be drawn through point P that will never intersect AB. That line and line AB are parallel; they can be extended any finite distance and will not meet. The axiom can be stated, "Parallel lines never meet." Euclid, writing in the fourth century B.C., made no attempt to prove this assertion—as an axiom it was assumed self-evident—but it stood in curious contrast to the four axioms that precede it in his *Elements*. They are pithy; it is relatively lengthy and indirect. And while Euclid, having set forth the other axioms, proceeded to put them to work building his theorems, he seemed wary of the fifth and invoked it relatively seldom. These and other clues led mathematicians to suspect that the axiom of parallels was not a true axiom at all but a theorem, derivable from the other four axioms. Countless mathematicians down through the centuries tried to derive it. All failed. In retrospect, their failure became a tribute to Euclid's genius in not attempting the feat himself. As the British mathematician Herbert Turnbull wrote, "There is dignity in the way that Euclid left this curious rugged excrescence, like a natural outcrop of rock in the plot of ground that otherwise had been so beautifully smoothed."

By the nineteenth century, enlightened by new developments in the field of symbolic logic, three geometers independently discovered that the reason the axiom of parallels could not be proved was because it was not necessarily so. That slim passway opened onto the non-euclidean geometries, one of the major intellectual revelations since Athenian times.

The German mathematician Karl Friedrich Gauss appears to have been the first to realize that strange new geometries could be constructed by denying Euclid's fifth axiom and substituting another in its place. The results were so weird that although Gauss enjoyed great prestige, and was widely regarded as the greatest mathematician of the time, he withheld his work for fear of ridicule.

In Hungary, János Bolyai was at work on the problem against the advice of his mathematician father, who wrote him, "I entreat you, leave the science of parallels alone. . . . I have traveled past all reefs of this infernal Dead Sea and have always come back with a broken mast and torn sail." In 1823 Bolyai discovered that by denying the parallels axiom he could create a new geometry, consistent as Euclid's but quite different. Nikolai Lobachevski in Russia made the same discovery at about the same time. Bolyai and Lobachevski's geometry was what is now called hyperbolic, meaning analogous to the three-dimensional form of a hyperbola, a shape that somewhat resembles a saddle. Bernhard Riemann of Göttingen then composed a spherical non-euclidean geometry. Riemann's Göttingen colleague, Felix Klein, incorporating work by Arthur Cayley of Britain, proved that these new geometries stood on equal footing with Euclid's. Logically, all were valid.

Traditionally science had viewed the natural world, however unruly it appeared, as embedded in a rigid three-dimensional geometry. A flower, an ocean wave or a star could always be mapped on the unchanging grid of left-right, up-down, forward-back. Now these new developments raised the odd possibility that three-dimensional geometry itself might be "curved," embedded in a lattice of four dimensions or even more. A four-dimensional substructure may or may not be visualizable—Bertrand Russell was convinced it was impossible for humans to imagine, say, a four-dimensional "supercube," while the physicist and physiologist Hermann von Helmholtz felt children might be taught to do so—but whether or not we can grasp such concepts as hyperdimensional geometry or "curved" space, we can think of ways to determine whether nature is built in accordance with them.

Imagine a race of two-dimensional creatures, able to see only frontward, backward and sideways, unable to perceive up and down, who live on the surface of a globe like the Earth. These flat earthlings wander the globe extensively and satisfy themselves it is boundless and has no edge. Then a flat Magellan circumnavigates it. They realize that although their home is unbounded it is not infinite. The flat geometers retire to deliberate. They hit upon the answer. We live, they announce, on the surface of something called a *sphere*. We cannot visualize such an object because it involves a third dimension in addition to the two we know, but here is the geometry of it, perfectly rational and consistent, like a circle but with a dimension added.

To convince flat skeptics, the flat geometers draw a large triangle and measure its angles. Bulging subtly with the unseen curve of the globe, the triangle adds up to a total of more than the 180 degrees of a flat triangle. The presence of an unseen dimension can be deduced though it cannot be perceived.

"There you are," they say. "We live on the surface of a three-dimensional object, a sphere, which to us appears finite but unbounded. Nature is built in more dimensions than we are accustomed to thinking of."

The creation of the non-euclidean geometries suggested that we three-dimensional humans might be in a situation something like that of the flat scientists. The universe may be built not in three dimensions but in four, or rather, as Mach would put it, may respond more readily to inquiry in four dimensions than in three. The idea that nature might be rooted in a substructure so different from the one we are accustomed to was at first almost frightening, which may help explain why Gauss did not care to see it loosed into the world during his lifetime. As it happened, the initial impact of the non-euclidean geometries was largely restricted to mathematicians, philosophers and a few scientists, because no evidence existed linking these abstractions to the real world. Four-dimensional universes might flourish in the pages of books, but Euclid and Newton still ruled Earth and sky. Or so it seemed until Einstein.

In Zurich and Berlin in the years 1912–16, Einstein worked on general relativity, his theory of gravitation. In addition to the philosophical outlook of Mach's Principle and the power of non-euclidean geometry, he made use of the field concept, which had originated in electromagnetics but which he intended to apply to gravitation. The results would do away with the Newtonian concept of gravity as a force acting at a distance, a notion Newton himself had admitted was inexplicable. "Action at a distance," Einstein wrote, "is replaced by the field."

Einstein's vision of the completed theory, at least in outline, may have been clear during these years—his intuitive sense of nature was unsurpassed—but working out the details proved difficult. Non-euclidean geometry is subtle and Einstein was not much of a mathematician. "Every boy in the streets of our mathematical Göttingen understands more about four-dimensional geometry than Einstein," wrote the geometer David Hilbert, with forgivable exaggeration. "Yet," Hilbert added, "despite that, Einstein did the work and not the mathematicians." Like Gauss, who once cheerfully told a colleague, "I have the result, only I do not yet know how to get *to* it," Einstein knew his goal but was not sure of the path leading there. Unlike Gauss, he had no confidence he would ever find it on his own.

His close friend Marcel Grossman was a devotee of the non-euclidean geometries. "Help me, Marcel, or I'll go crazy!" Einstein wrote. Under Grossman's tutelage, Einstein gained respect for Hermann Minkowski, whom he remembered from his student days as a cold lecturer who gave him low marks, but now recognized to have been a pioneer in a central idea of general relativity—that non-euclidean geometry could be applied to the physical world by designating time a fourth dimension. When relativity burst upon the world, Minkowski remarked that "for me it came as a tremendous surprise, for in his student days Einstein had been a lazy dog. He never bothered about mathematics at all." Grossman tutored Einstein in the tensor calculus of Ricci and Levi-Civita, the intricate machinery of the quadratic differential equations, and the incisive analysis of non-euclidean geometry published sixty years before by Bernhard Riemann (whose death of respiratory illness before his fortieth birthday left forever unanswered the question of how close he might have come to relativity himself). "Only the genius of Riemann," Einstein later wrote, "solitary and uncomprehended, had already won its way by the middle of the last century to a new conception of space, in which space was deprived of its rigidity, and in which its power to take part in physical events was recognized as possible."

Newton viewed space and time as separate and absolute. As conceived by Einstein they are united in a flexible continuum that responds to the presence of matter. The stars and planets wrap the spacetime continuum around themselves, in a sense, each sitting in the center of a sort of spacetime whirlpool. The commerce we call gravity occurs because objects follow the easiest, most efficient course over the undulations of the continuum. Earth in its orbit glides along inside the sun's spacetime vortex like a roulette ball whirling above the wheel, balancing its velocity against its tendency to slide toward the sun. That tendency to fall "down" the slope of spacetime is equivalent to gravity, but no "force" of gravity is postulated. Light beams also follow the dips and hills of the continuum. They trace trajectories we call "bent," though that is just three-dimensional parochialism talking; they are going just as straight as the shape of spacetime permits.

Einstein's accomplishment, of course, was not to propose that space is like a whirlpool and the planets roulette balls, but to create a theory that explained the behavior of heavenly bodies more accurately and coherently than had its predecessors. General relativity accounted for a peculiarity in the long-term orbital behavior of the planet Mercury that had been inexplicable in Newton's theory. It predicted that starlight passing near the sun would appear, to us in three-space, to be bent, to a degree verified by Eddington in an historic solar eclipse observation on Principe Island off Nigeria on May 29, 1919. And general relativity has passed every subsequent experimental test to which it has been put.

Relativity was fresh air to cosmology. So long as space and time had been regarded as inert and unchanging, merely a stage upon which matter played out the cosmic drama, there had seemed to be no escape from the equally illogical stances of assuming the universe finite or assuming it infinite. But now relativity showed that in the oceanic reaches between galaxies, space and time could be so distorted that the universe itself might be said to have a shape. Suppose it were a four-dimensional sphere. Such a universe is finite—it contains a finite number of galaxies—but unbounded: an immortal cosmonaut can circumnavigate the universe and never reach a boundary, though eventually he will have visited every galaxy. Observers in all the galaxies, looking along the gently curving light beams, see galaxies in every direction; there is no edge in space to which Lucretius' spear-thrower can book passage to perform his experiment.* So the paradox is broken. "This suggestion of a finite but unbounded space," exulted the physicist Max Born, "is one of the greatest ideas about the nature of the world which has ever been conceived."

Many other uniform four-dimensional shapes of the

---

* If the universe was created at a certain time in the finite past, however, we may consider that beginning an "edge" in time. That in fact is what most astronomers now mean when they speak of an edge of the universe. See Chapter 8.

universe seem to be possible. The models fall into three categories, "closed," like the finite but unbounded sphere just described, "open," or hyperbolic, and "flat," meaning euclidean, a middle case between the other two. The real universe presumably can conform to only one of the three. The question is, which one?

In 1916, when general relativity was published, to determine such a thing as the shape of the cosmos was a task beyond astronomy, which still was uncertain even if there was anything beyond our own galaxy. Cosmologists busied themselves studying not the sky but the ramifications of Einstein's theory. This was pleasant work—watching the mathematics of relativity unfold is comparable to listening to a Bach fugue—and it led to a startling consequence.

Relativity implied that the universe expands.

Einstein caught sight of this early but found it unsettling even for him. That the universe should be expanding was, as the physicist John Wheeler wrote years later, "a prediction too fantastic for its author to credit." Every astronomer Einstein talked to told him the cosmos was static and unchanging, but when Einstein tried to create a cosmology based upon general relativity that would conform to this view, the theory resisted his efforts. It hinted that the universe must not be still.

For once in his career Einstein lost confidence and balked. To make relativity deliver the static universe that he had been told fit reality, he introduced a new term into the equations that he called the "cosmological constant," designated by the Greek letter lambda. It represented a sort of antigravity operating over long distances. Its effect was to make possible a relativistic model of the universe that would neither expand nor collapse. Einstein published a cosmology incorporating it, and soon was sorry. Within two years he had come to regard the cosmological constant as "gravely detrimental to the formal beauty of the theory." Three years later he volunteered that it "constitutes a complication of the theory, which seriously reduces its logical simplicity." By 1931 he discarded it altogether, branding it the worst blunder of his career.

Meanwhile cosmologists across Europe applied the powerful new tool of relativity with less caution. In clusters of chalked symbols, the cosmos bloomed like a flower, danced like a trained bear, warped and curved in an eerie calculus beyond the eye. But it was a cosmos only of the mind. Few expected that reality could be so wild.

Willem de Sitter, director of the Leiden Observatory in the Netherlands, began speculating on new cosmologies almost from the day a copy of relativity theory arrived in the mail from Einstein. An astronomer in the traditional mold, with a white beard and an absentminded manner, De Sitter had a supple and inventive mind. For simplicity, he based his calculations on the assumption that the universe was completely empty. This approach might seem questionable, since we know the real universe does contain matter, including ourselves, but De Sitter meant his model to be an approximation of reality, a mathematical fiction that might have something interesting to say. After all, he would say, the cosmos is mostly space anyway, so what is wrong with a cosmology that is all space? The De Sitter universe was an expanding one so far as its geometry was concerned, but as De Sitter remarked with a grin, it could also be considered static, because there was nothing in it to expand.

Eddington introduced particles of matter into this vacant universe and found that they tended to fly apart, but De Sitter at first did not take the implication very seriously. He was preoccupied with the much-debated "De Sitter effect," a mathematical oddity that made distant objects in his model look as if they were receding even if they weren't. This effect was later discredited so far as the real world is concerned, but meantime it caused confusion for a number of researchers, including Hubble.

The first cosmology to suggest clearly that the universe is expanding came in 1922. Its author was Alexander Friedmann, a Russian mathematician who watched the growing clamor over relativity from a position almost on the sidelines, in Petrograd. Like the other cosmologists, Friedmann had no new observational evidence concerning the behavior of the real universe; his technique, like theirs, was to pore over the mathematics of relativity and see whether there was a previously unnoticed lever to be pulled or gear that could be engaged. He was rewarded when he discovered that Einstein, in his "cosmological constant" model, had made an algebraic error, essentially dividing by a quantity that could in some cases be zero. When Friedmann corrected this error, the Einstein universe promptly ceased to be static and glided into motion. Friedmann found that depending upon starting conditions, the universe might expand, or contract, or even oscillate, expanding for eons, then collapsing. Relativity, defying even Einstein's efforts to restrain it, implied a dynamic cosmos.

Whether the real cosmos acted accordingly was a question only observation could answer, but observers, generally speaking, knew little cosmology, and many cosmologists knew little astronomy. The big new telescopes were in America, while cosmology was still primarily European, its practitioners attached to universities at

Leiden, Berlin or Cambridge. An indication of the gulf separating the two fields was the lack of attention paid Slipher's work at Lowell Observatory on the shift of spectral lines in light from spiral nebulae. The year Friedmann published the first of his studies of cosmological relativity, Slipher accumulated, though he had not yet published, a list of forty galaxies for which all but four displayed redshifts, potent if preliminary evidence that the universe was expanding. The list could have aroused the greatest interest in Einstein, Friedmann, or any of the cosmologists conversant with Friedmann's work, but apparently none of them heard of it. Slipher, it is true, was employed by an unusually isolated observatory, whose general secretary responded to official inquiries from outside astronomers by throwing them away. But all American astronomers were, to some degree, detached from the revolution in theoretical cosmology going on across the sea. For want of hard astronomical evidence, Friedmann's expanding universe was dismissed, by those few who bothered to read it, as little more than a mathematical curiosity. Einstein acknowledged it briefly in a polite but unenthusiastic letter. Friedmann died, of pneumonia contracted on a cold-weather excursion in a meteorological balloon, too soon to know the importance of his achievement.

After Friedmann's death his theory sank from view with so few ripples that when a young Belgian researcher approached the same problem only five years later he came across no trace of it and was obliged, unwittingly and laboriously, to duplicate many of Friedmann's calculations, arriving at almost the same results. This was the Abbé Georges Lemaître, professor of relativity and the history of science at Louvain. Like Friedmann, he found in relativity the seeds of universes expanding, contracting or oscillating. But his carefully thought-out cosmologies were published in a journal not widely read (the tributary *Annals of the Brussels Scientific Society*) and, like Friedmann's, they descended into obscurity.

Lemaître anticipated that the universe would be found expanding and that this could be confirmed by looking for redshifts in the spectra of galaxies. Probably he had read Slipher's table of redshifts, which Eddington, who had been his teacher, had just published.

The year was 1927, one of the last before cosmologists and astronomers began talking to one another in earnest. A third of the way around the world from Louvain, at Mt. Wilson, Edwin Hubble was sifting from his data a portrait of galaxies rushing from one another. Later, when his discovery linked the two disciplines, he would write a book titled *The Observational Approach to Cosmology*, but at the time he knew almost no cosmology. He knew nothing of the work of Friedmann and Lemaître. He had read De Sitter, but only enough so that at first he attributed his redshifts of galaxies to the spurious "De Sitter effect."

With Hubble's discovery of the redshift-distance relationship, and therefore of the expansion of the universe, "Einstein's theory of general relativity came into its own as a physical theory," the latter-day cosmologist William McCrea wrote. "Hitherto it had predicted minute departures from classical physics. Now it had successfully made the greatest prediction in the history of science." But Hubble did not know this when he published his milestone 1929 paper.

Hubble learned of Lemaître's cosmology in 1930 by reading about it in a semipopular article of Eddington's, and news of the expansion of the universe reached the man who had seen it in the sky. Alone in his office Hubble may have danced a jig for all we know today, but he was stone-faced as ever in public. For several years he was reluctant even to admit his observations evidenced an expanding universe. In 1937 he conceded grudgingly, "Well, perhaps the nebulae are all receding in this peculiar manner. But the notion is rather startling."

4

LOOKBACK TIME

. . . all the bells that ever rang still ringing in the long dying light-rays. . . .

—WILLIAM FAULKNER

A PROBLEM AFFLICTED HUBBLE'S MODEL OF THE COSMOS. The numbers were wrong. First, all the other galaxies appeared to be smaller than ours. M33, for example, measured only one-twentieth the size of the Milky Way, the Andromeda Galaxy perhaps one-fifth. Either the figures were off or else the universe consisted of a lot of small galaxies grouped around our galaxy, little dogs ringing a big dog, a situation that would award us suspicious eminence. Second, Hubble found globular clusters arrayed in a halo around the Andromeda Galaxy, as Shapley had found them around ours, and his calculations demanded that either the Andromeda Galaxy was farther away than thought or else its globular clusters were, for some reason, intrinsically fainter than ours. Third, the Hubble value for the rate of expansion of the universe, if run backward in time, mashed the cosmos to high density at a time more recent in history than what geologists said was the age of Earth. These three discrepancies—the relative dwarfing of galaxies, the dimness of their globular clusters, and the unseemly youth of the cosmos—flawed Hubble's work for years, though to his credit he made no attempt to paper them over. Various theories were proposed to explain the redshifts as caused by something other than velocity, but all were unconvincing, and Hubble, though he flirted briefly with a few of them, refused to be seduced. De Sitter shrugged off the paradox whimsically— "After all," he wrote, "the 'universe' is an hypothesis, like the atom, and must be allowed the freedom to have properties and to do things which would be contradictory and impossible for a finite material structure"—but Hubble declined that explanation as well. "We face a rather serious dilemma," he wrote. "Some there are who stoutly maintain that the earth may well be older than the expansion of the universe. Others suggest that in those crowded, jostling yesterdays, the rhythm of events was faster than the rhythm of the spacious universe today; evolution then proceeded space, and, into the faint surviving traces, we now misread the evidence of a great antiquity. Our knowledge is too meagre to estimate the value of such speculations, but they sound like special pleading, like forced solutions of the difficulty."

It took three decades to set the numbers right, because they were infected by several different errors immediately resolvable by neither theory nor observation. Exposing them required a comprehension of how stars work—that came with the collaboration of physicists and astronomers in the new discipline of astrophysics—and construction of a much larger telescope. The central figure in both developments was George Ellery Hale.

When Einstein and Hubble met in 1931 at Mt. Wilson, the walls dividing theoretical physics from astronomy were finally crumbling. Hale supervised that demolition. He insisted that the three giant observatories he built—Yerkes, Mt. Wilson and Mt. Palomar—be equipped with the spectrographs, darkrooms and machine shops needed to make them physical laboratories where the stars could be analyzed as well as observed. To publish the results he founded and edited the *Astro-Physical Journal,* which lost the hyphen in its title as the alliance took. The style of modern astronomy, an aggressive probing, almost a dissection of the heavens, comes from Hale more than from any other man.

It developed from Hale's love for the sun. As a child he learned that sunlight could be taken apart for analysis, a discovery from which he never recovered. He pored over charts of the solar spectrum with the excitement of Columbus scrutinizing Pierre d'Ailly's *Imago Mundi.* At age twenty he invented the first successful spectroheliograph, an instrument that photographs the sun in the light of a single slice of its spectrum. In some wavelengths the sun looked as convoluted as a living cell. In others, it resembled a balloon flecked with ivory speckles. The spectrohelio*scope,* a device Hale built years later, allowed direct and continuous viewing of the sun in these exotic costumes.

Hale had a strong sense that we live *in* the universe, and can learn about it by studying things nearby as well as distant. "The sun is a star, the only star whose phenomena can be studied in detail," he told contributors to his ceaseless building projects. He had read Darwin and was convinced that physics and astronomy, in cooperation, could trace the evolution of stars as Darwin had onions and cows. In 1908 Hale wrote a prophetic sentence perhaps a century ahead of its time: "We are now in a position to regard the study of evolution as that of a single great problem, beginning with the origin of stars in the nebulae and culminating in those difficult and complex sciences that endeavor to account, not merely for the phenomena of life, but for the laws which control a society composed of human beings."

His career as a telescope builder began one twilight on the veranda of the Powers Hotel in Rochester, New York. Relaxing among colleagues after reading a scientific paper to a gathering inside, he overheard Alvan Clark, a grinder of fine lenses, remark that the University of Southern California had ordered objective lenses for a 40-inch telescope made by a Paris optical shop and then had been unable to pay for them. The two slabs of cast glass, which if ground to figure and placed together would form the lens of an astronomical refracting telescope, were in storage in Paris. Hale was interested. A 40-inch would be the world's largest telescope, surpassing the 36-inch at Lick Observatory. Considerable money would be needed to buy the blanks, have them ground, and build an observatory around them, but Hale, the son of a wealthy father, felt he might be able to raise it. Back in Chicago he made the rounds of potential philanthropists, endured a string of polite refusals, then met and won over Charles Yerkes. Years of further haggling and pleading followed, but the recalcitrant Yerkes finally delivered, in bits and pieces, all he had pledged, and Yerkes Observatory was dedicated October 21, 1897, with Hale, age twenty-nine, its director.

The observatory at Williams Bay, Wisconsin, had hardly begun operations before Hale was camped in an abandoned shack atop Mt. Wilson, testing the night skies with a small telescope. Convinced the site was suitable for astronomy, he gathered backing for a solar observatory there. The big deep-space telescopes followed, beginning with the 60-inch, its mirror a gift from his father.

Though Hale was physically healthy—he ascended Mt. Wilson at a sort of loping run—he suffered nervous disorders aggravated by overwork, by byzantine negotiations with millionaires and their foundations, and by fear that his ambitious projects might fail. When the 100-inch telescope was completed November 7, 1917, Hale looked into the eyepiece and saw, instead of a star, a blurred mass of overlapping discs. As it turned out, the mirror was fine. It had been only temporarily distorted that afternoon when a workman let sunlight fall through the dome slit on it, expanding its thick glass. But Hale, waiting sleepless in his bunk until 2:30 A.M., when the mirror regained its figure, went through one of many traumas that punctuated his career. By middle age he was beset by insomnia, headaches, numbness in the feet, an inability to read. An imaginary little man appeared to keep him company on sleepless nights. Hale, chatting with this amiable elf, feared he was going insane.

Under doctors' orders he resigned the directorship of Mt. Wilson in 1922 and attempted to retire. He built a small solar observatory near his home in Pasadena, equipping it with a machine shop and a 12-inch lens he had used as a student. There he invented the spectrohelioscope, outlined the magnetic properties of sunspots, and predicted radio storms on Earth set off by eruptions on the sun. He wrote books and articles. He tried to find money to build an observatory in the southern hemisphere. Soon he was back on the psychiatrist's couch.

By then the public's attitude toward science was warming, and Hale's victories began to come more easily. The accomplishments of Hubble, Shapley and others under Hale's direction had aroused American curiosity and pride, while enthusiasm for European scholarship cooled. Hale himself, who fumed through a year of graduate study at the Physikalisches Institut in Berlin while approving blueprints for the Yerkes Observatory he would return to direct, belonged to perhaps the last generation of American scientists who considered it obligatory to round out their educations in the Old World, and Hitler would soon lend his support to the westward shift of the scientific epicenter. In 1932 Einstein and his wife packed for what she thought would be only a brief trip to California. Einstein asked her to take a good look at their villa, because "you will never see it again." Their trip became permanent exile. Nazi stormtroopers sacked the house.

Hale found a receptive audience for an article proposing

a new large telescope. "Starlight is falling on every square mile of the earth's surface," he wrote, "and the best we can do is to gather up and concentrate the rays that strike an area 100 inches in diameter." A large telescope would be expensive, of course, but "if the cost of gathering celestial treasure exceeds that of searching for the buried chests of a Morgan or a Flint, the expectation of rich return is surely greater. . . ." The price of a 200-inch telescope on Mt. Palomar would be six million dollars. With a readiness that surprised Hale, the Rockefeller Foundation pledged the money.

The 16-foot, 8-inch mirror was cast by the Corning Glass Company in New York. In working their way up to it, the glassmakers produced a whole generation of mirrors for new telescopes. Two 61-inch blanks went to Shapley at Harvard, a 76-inch to Toronto, an 82-inch to McDonald Observatory in Texas, a 98-inch to the University of Michigan, and a 120-inch, eventually, to Lick. Poured in liquid pyrex, with a honeycomb back to minimize weight, the 200-inch mirror cooled for a year in a sealed brick igloo. The glassmakers sandbagged the Corning plant against a river flood to stop rising water that could chill the igloo and crack the mirror. Schoolchildren across the country watched the finished blank go by, set on edge in a steel case aboard a custom-built railway car drawn at twenty-five miles per hour. In Pasadena, opticians clad in white smocks ground the blank for two years, were interrupted by the war, then ground two years more, removing five and a quarter tons of glass to figure the mirror into a parabola that, when coated with a layer of aluminum 1,000 molecules thick, would bounce starlight to a clear, flat focus. The opticians figured the mirror to within two millionths of an inch, an accuracy finer than the vagaries of Earth's atmosphere would ever permit the telescope to realize in practice. It was a matter of pride with them. The dome for the 200-inch and the smaller domes surrounding it were erected in a pasture at 5,500 feet altitude, where tests indicated that the clarity and steadiness of the air, what astronomers call "seeing," were optimal.*

Hale, supervising the 200-inch project, cautioned that he might not live to see it completed. By 1936 his health was failing and he could no longer make the trip up Mt. Wilson. Many astronomers there had never met him and regarded him, a biographer wrote, as "almost a mythical being." Invited by Shapley to attend a symposium in his honor, Hale declined in a note that read, "Old and battered fossils may retain a certain antiquarian interest, but in the midst of recent revolutionary advances, they are rapidly outclassed." By early 1938 he could no longer use the small solar observatory he had built for his so-called retirement. He wanted to visit Palomar and see the big dome under construction but had to keep postponing the excursion for reasons of health. He died February 21.

Throughout Hale's career the astrophysics he had nourished grew. The story of how stars evolve was being pieced together in an international effort that many astronomers

---

* Observatories are built on mountains to raise them above the thick, turbulent lower levels of the atmosphere. In principle, the higher the telescope the better. In practice, the expense of building mountain roads and the severity of high-altitude weather inspires a willingness to compromise. Mauna Kea Observatory in Hawaii, which may be the highest observatory that will ever be built on Earth, stands at nearly 14,000 feet, where gale-force winds are common and the air so thin that higher brain functions are impaired by lack of oxygen. Astronomers quartered halfway down the mountain write themselves childishly simple instructions they hope their muddled brains will be able to obey when they go up to the dome to observe.

and physicists consider the greatest achievement of both their sciences.

The early chapters of this story were written just after the turn of the century, when Shapley's Princeton mentor, Henry Norris Russell, and a Danish chemical engineer, Ejnar Hertzsprung, independently discovered a relationship between the colors of stars and their absolute magnitudes. Further research over the years elaborated the Hertzsprung-Russell diagram into a compelling outline of the physical society of stars in which average stars belong to a gently curving "main sequence," while dwarfs, giants and supergiants occupy branches to either side. The pattern persists when we take into account the color, mass and chemical composition of the many members of the family of stars.

With the rise of nuclear physics came some understanding of what goes on inside stars and why they should fall into the graceful, tree-trunk shape of the Hertzsprung-Russell diagram. The operative mechanism is nuclear fusion, energy released when particles in the nuclei of atoms recombine into configurations of slightly less mass than they started out with. Einstein's famous equation $E = mc^2$—energy equals mass times the speed of light squared—governs the transaction. Because the speed of light is a large number, 186,000 miles per second, even a tiny amount of mass converts into a great deal of energy. Research dating from 1939, when the physicist Hans Bethe first proposed that nuclear fusion was responsible for the furious heat of stars, has established that a variety of fusion systems probably operate within stars at various stages of stellar lifetimes, some of them simultaneously.

As pieced together by astronomers and nuclear physicists, the life story of a star goes something like this: A star forms by gravitational contraction of dust and gas and begins to glow when temperature and density at the core rise sufficiently for fusion to operate. It enters the main sequence on the Hertzsprung-Russell diagram and settles down there for about 90 percent of its lifetime. That is where the sun is now. Ultimately the star depletes the hydrogen fuel it has been processing at the core, and must begin to rely more heavily upon fusion of helium nuclei to survive. Its outer atmosphere expands into a dull red cloud; on the Hertzsprung-Russell diagram it leaves the main sequence and skids into the realm of the giants. When all fusion mechanisms at the core fail for want of fuel, radiation pressure from within can no longer balance the grip of gravity, and the star collapses. The length of time this tale consumes for any given star depends upon how much material went into forming it: A humble yellow star like the sun stays on the main sequence glowing steadily for billions of years before skirting through a phase as a modest red giant, then becoming a white dwarf. Larger blue-white stars may squander energy, remain on the main sequence for a much shorter time, balloon into enormous giants, and then, theatrical in death as in life, crush themselves through the dwarf phase to become crazily spinning neutron stars or perhaps black holes.

Though the full story of stellar evolution is far from being understood even today, its outlines were sketched sufficiently by the 1930s to be interesting to cosmologists. If pondering the geometry of the cosmos was like mapping the streets of a city, studying about the mechanisms of stars was like learning about the city's inhabitants. A pioneer in applying the new knowledge of the evolution of stars to cosmology was Walter Baade.

Baade was a student in Germany when the Hertzsprung-

Russell diagram was first published. He recalled that it "made a tremendous impression on me, because obviously it opened up an entirely new way of studying galactic structure. We could pick out well-defined types of stars . . . and use them to analyze the structure of the galaxy." He would apply discoveries in stellar evolution to understanding not only the structure of our galaxy but the scale of the universe.

When Baade arrived at Mt. Wilson in 1931, having emigrated from Germany shortly before Einstein did, he guessed, with impeccable intuition, that the problems astronomers were having trying to estimate the age and size of the universe had to do with insufficient understanding of their stars. Stars, after all, specifically variable stars, were the means Shapley and Hubble used to determine cosmic distances.

Earlier, Shapley and Robert Trumpler of Lick had noticed that stars clustered in our galaxy tended to be idiosyncratic to their home cluster; that is, they generally resembled each other, in terms of their place on the Hertzsprung-Russell diagram, more closely than they did stars in other clusters elsewhere in the galaxy. This implied, at least in retrospect, that stars within a galaxy are born at differing times as the galaxy evolves, and that there might exist distinct generations of stars, each with its own peculiarities. Baade chose to investigate the question by examining not our galaxy but another similar one, a spiral both large enough to resemble our own and nearby enough so that its individual stars could be seen. The obvious candidate was M31 in Andromeda. M31 had once been a passion of Hubble's, but he had since moved on to observe deeper space, and Baade, five years Hubble's junior and a

Albert Einstein in California, 1933. Einstein's visits there in the early 1930s brought him face to face with Hubble, Humason, and other astronomers whose observations of the expanding universe fulfilled one of the strangest prophecies of relativity theory.

George Ellery Hale reconnoiters Mt. Wilson. The telescopes he built on its peak extended human vision of the cosmos by a factor of a thousand.

Hale (second from left) poses in Pasadena with a group of scientists and financiers including the naturalist and explorer John Muir (center front) and Hale's benefactor, Andrew Carnegie (far right).

A section of the tube for Hale's 60-inch telescope is readied for transport from Pasadena up the gravel road to Mt. Wilson, circa 1915.

From 1914 to 1921, Harlow Shapley charted the outlines of our galaxy, discovering that it resembled an enormous pinwheel with the sun located about two-thirds of the way out toward its rim. His departure from Mt. Wilson at age thirty-six deprived Hale of one of his most brilliant astronomers.

HARVARD
COLLEGE
OBSERVATORY

Dedication of the 200-inch telescope on Mt. Palomar. Starlight is collected by the mirror at the base of the skeletal tube (in this photograph the mirror is hidden beneath steel off-duty shields), then reflected to a focus at the observer's cage suspended in the top third of the tube. For extreme deep-space work—what Hubble called "observational cosmology"— the observer rides inside the cage.

Edwin Hubble, discoverer of the expansion of the universe, in the observer's cage of the 200-inch telescope. Mirror is below and to the left.

The spiral in Andromeda, M31, closely resembles our Milky Way, though it is somewhat larger: Astronomers there, if there are any, may find our galaxy as lovely a spectacle as we find theirs. The bright patches above and below the Andromeda spiral are satellite galaxies, comparable to our Magellanic Clouds. The many individual points of light are foreground stars in our galaxy.

newcomer to the observatory, was happy to find he could work on M31 all he wanted without crowding Hubble.

With the help of Joel Stebbins, a Wisconsin astronomer who frequently observed at Mt. Wilson, Baade first refined to new accuracy the zero-point used in measuring star brightnesses on all photographs. This corrected Hubble's figures slightly upward, though not nearly enough to resolve the dilemma of a universe seemingly younger than the earth. Baade went on to study the stars of M31, an adventure that eventually consumed twenty years. Resolving stars near the edges of the galaxy's spiral arms was not difficult—Hubble had accomplished this before—but in many ways the most interesting part of a galaxy is its densely populated center, and here, though calculations indicated the 100-inch was up to the task, Baade was unable to photograph individual stars. He tried for years but could get nothing better on his plates than the sort of grainy porridge that had frustrated Hubble in the 60-inch days.

When the United States entered World War II most of the Mt. Wilson astronomers were diverted to wartime duty. Hubble was shuttled off to direct a ballistics project in Aberdeen, Maryland, grumbling haughtily that he didn't know what "ballistics" meant. But Baade, born and raised in Germany, had lost his U. S. citizenship papers, and found himself classed an enemy alien. Barred from security clearance, he was left alone with the 100-inch telescope. As if in an astronomer's fantasy, he watched from Mt. Wilson as the lights of Los Angeles disappeared into wartime blackout. With the skies darker than he had ever seen them, he again attempted to resolve the nucleus of the Andromeda Galaxy into stars. Again he failed.

As a last resort Baade switched from the customary photographic emulsion, which had a peak sensitivity near the blue end of the spectrum, to red-sensitive plates, which were less vulnerable to contamination from natural background skylight.* The drawback of red-sensitive plates was that they were very "slow," requiring exposures of eight hours or more.

A lot can happen to a large telescope in eight hours. The figure of the mirror changes minutely as temperature drops during the night. The girders of the tube contract and alter the focal length. The quality of the air varies moment by moment, blurring star images cumulatively as the exposure goes on. To obtain pinpoint images of stars in M31, Baade charted the weather and scheduled his sessions on nights when seeing was likely to be best, a luxury attainable because he had the telescope almost to himself. On afternoons before observing runs he ordered the dome opened and the slit turned away from the sun, so that the temperature inside would more nearly equal that outside. When full darkness came he aimed the telescope at the center of the Andromeda Galaxy, opened the plate, and guided on a star magnified 2,800 times. Using a technique he invented, he learned to monitor this blob of light for signs of disturbances in the air, quickly changing the focus to compensate for them.

"I still remember how confused I was as to what to do," Baade said, recalling the first night he tried his new method. "But you just calm down and wait for a moment of good seeing; and after you have mastered it, it is astonishing how little time you need to see what the situation is, and to give a quick turn to the focus. By

---

* Human light pollution aside, the sky glows with scattered starlight, faint aurora, and zodiacal light, which is sunlight reflected from interplanetary dust.

training myself in this way I finally learned to handle the method. . . . After shooting was over, it was quite clear that all the precautions had actually been necessary; I had just managed to get under the wire, with nothing to spare." The photographs showed thousands of individual stars in the heart of M31. The margin was so slim that the *Astrophysical Journal* could not trust the printing process to show that the stars had been resolved; the journal had to make individual enlargements from the original negatives and bind them into each issue with Baade's paper.

The interesting thing about Baade's achievement was its strong indication that the stars in the nucleus of M31 were themselves predominantly red. This was reinforced when Baade tried to resolve the nucleus with blue-sensitive plates again, employing the same rigorous techniques he had with the red plates, and again failed. The results fit well with his growing suspicion that the stars in a normal galaxy were broadly divided into two groups. One group, now understood to be older stars, he called Population II; they were red or yellow giants, lacked atoms of heavy metals, and were found primarily in the galactic nucleus and in globular clusters. Population I stars, now known to be younger, were yellow or blue, and contained a variety of heavy elements that had been seeded into the interstellar medium by their Population II ancestors. They were found along the galaxy's spiral arms; the sun is a Population I star.

What if variable stars also came in two varieties? When Shapley charted interstellar distances by using Cepheid variables, he assumed that the two sorts of variables he employed were essentially alike. Some were short-term, oscillating in brightness within a matter of hours. Others varied over periods of weeks or months. Each obeyed the

period-luminosity relationship discovered by Henrietta
Swan Leavitt—the brighter they were, the longer their
period. But only short-term variables were close enough to
our sun for Shapley to have measured their distances with
any precision. He had extrapolated from these, drawing a
curve up through the data points for the long-period
variables, and that curve was the basis for both his distance
calculations and those of Hubble. In 1944 Baade proposed
that long-term and short-term variables were actually quite
different sorts of stars. He predicted that the long-term
variables would prove to be Population II stars, intrinsically much brighter than their short-term, Population I
nephews. If brighter, then to look as dim as they did, they
must be farther away. If Baade was right, the galaxies were
farther apart than in Hubble's figures, so the universe had
been expanding for a longer time and the time-scale
problem would be solved.

Resolution of the question awaited completion of the
200-inch telescope. Shapley predicted that short-term variables in the Andromeda spiral were bright enough to be
seen with the new telescope. Baade predicted that even the
200-inch would fail to find short-term variables there,
because the Andromeda Galaxy was too far away.

The Palomar telescope with its 15-ton mirror was ready
for preliminary tests by December 1947. For first use, in a
ceremony known as "letting in the light," it was rigged for
visual observation, a rarity for large instruments, where
photography and spectroscopy take precedence and observers seldom *look* at anything for long. A little hand spyglass
was mounted for use as an eyepiece. John Anderson, who
was Palomar project director—he had camped on the
mountain on site-selection trips with Hale twenty years

before, the two men testing the skies all night while distant campfires flickered on the Pala Indian reservation below— was now the first to look through the 200-inch.

"What did you see?" he was asked.

"Oh, some stars," Anderson said.

For the benefit of guests the huge telescope was aimed at Saturn, but seeing was bad and the image watery. A globular cluster near the zenith was more of a success: Its stars stood against the black sky like jewel chips in ink.

Hubble waited out weeks of thunderstorms and bad seeing and then, a few hours before dawn January 31, rode the small elevator up to the observer's cage to test the telescope alone. Over the intercom he gave the night assistant the coordinates for Coma Berenices, a good test subject that includes a spray of dim stars in the foreground and a cluster of galaxies in the deep background. Hubble braced himself in the swiveling observer's chair as the telescope heeled over. When the coordinate dials told him he was on target, he focused, checked the field, and exposed a plate.

The mirror was still in rough condition, its outer 18 inches left unfinished until opticians could make their final tests, its reflective coating filmed with grime, but the first photographs told Hubble the telescope was going to work. Exposures of five or ten minutes recorded stars the 100-inch could barely reach in a full night. "The tests show quite clearly that the 200-inch opens to exploration a volume of space about eight times greater than that previously accessible for study," he reported. "The region of space that we can now observe is so substantial that it may be a fair sample of the universe as a whole."

When the mirror had been given its final figure, recoated and installed, Baade got his chance to study the Andromeda Galaxy. If Shapley were right and the galaxy was only 750,000 light-years away, the 200-inch should be able to photograph short-term variables there. But Baade's photographs showed no short-term variables.

Baade withheld his results until he could be sure the telescope really was performing to specifications. William Baum, an expert in photocell technology brought to Palomar to gauge the parameters of the new instrument, confirmed this to Baade's satisfaction by autumn 1952. Now Baade was ready, and at the next conference he attended, a meeting of the International Astronomical Union in Rome, he announced support for his theory that there are two kinds of Cepheid variable stars, that Shapley had confused the two, and that consequently the universe as charted by the Cepheid beacons was larger than had been thought. "The error must be such that our previous estimates of extragalactic distances . . . were too small by as much as a factor of two," Baade said. "Many notable implications followed immediately from the corrected distances: the globular clusters in M31 and in our own galaxy now come out to have closely similar luminosities; and our galaxy may now come out to be somewhat smaller than M31." When Baade finished speaking, a South African astronomer, A. D. Thackeray, stood and said that recent observations he had made of the Magellanic Clouds, a prime subject of the southern hemisphere observatories, indicated a dual nature for the Cepheids very much like that put forth by Baade.*

---

* Baade's division of stars into two populations has since been extensively abridged and subdivided, to the point that some astronomers would like to abandon it altogether. But his central point has survived intact. The nuclei of normal galaxies are made up wholly or predominantly of old Population II stars formed from almost pure primordial hydrogen and helium. These stars cooked heavier elements in their cores, it seems, later venting them off into space as they

Subsequent revisions of the distance scale, primarily by Allan Sandage (who found that Hubble had sometimes mistaken bright gaseous nebulae in galaxies for giant stars) fixed the distance to the Andromeda spiral at 2.2 million light-years. Estimates of all other intergalactic distances were increased proportionately. In the reconciled Hubble cosmos, the Milky Way was cast as a large but unextraordinary spiral galaxy, its stars and globular clusters no different from those in other systems. The expansion of the universe appeared to have been going on for at least 10 billion and perhaps as long as 20 billion years, a value in accord with the age astrophysicists assigned the sun, 5 billion years, and geologists the earth, about 4.5 billion. The universe at last seemed large and old enough for everybody.

For Hubble, relief that the paradox had been resolved was mingled with disappointment that he had not done it himself. "He was upset with the thing," a colleague said, "and he steered away from commenting on it, but he was not surprised. He always called everything he did a reconnaissance."

The expanded distance scale and the advent of the 200-inch telescope prompted astronomers to begin thinking in terms of photographing remote galaxies as they were appreciably long times ago. This was made possible, indeed obligatory, by the fact that the velocity of light is finite.

Imagine light from a distant galaxy traveling a billion light years and then encountering Earth. It rains down through the atmosphere, a sudden jolt after a billion years

collapsed. Subsequently new stars and planets were formed, incorporating the heavier elements, so that the metals so important to our civilizations were willed us by Population II stars like those Baade discovered.

of tranquil voyaging, and is absorbed by our pastures and forests, snowscapes, rooftops and seas, a minute addition to the plus side of the planet's energy ledger. One night a fraction of the light is gathered by the curved mirror of a telescope and recorded on a photographic plate. An astronomer develops the plate and looks at it. What does it show? A galaxy, certainly, but not a present-day galaxy. The light, after all, is a billion years old. Astronomers are cosmic historians, their telescopes time machines, their freshly developed plates relics ten thousand times older than the pyramids of Egypt. The term for this phenomenon is lookback time: Looking out is looking back.

When dealing with the large lookback times that pertain between clusters of galaxies, the observer finds it almost impossible to agree what the "distance" to a remote galaxy "is" within the meaning those words have for us on Earth. Our galaxy and the one we see on a given photographic plate were a certain distance apart when the light set out, but during the eons that passed before the light reached us, expansion of the universe increased the distance considerably. We may with equal justice say that the number of light-years separating us from the remote galaxy is equal to, less than, or more than the lookback time. Which answer we choose is as much a matter of philosophy as of science; the concept of distance has only limited usefulness over the cosmological sweep. As a result, the people who deal with the largest distances ever known seldom talk of "distance" at all, preferring to deal with unambiguous observables like a galaxy's redshift, apparent brightness, and how much of the sky it takes up.

The prospect of photographing galaxies at large lookback times had enormous potential interest to astronomers. It offered a way to study cosmic history directly. If the universe once expanded much faster than it does today, that fact ought to be ascertainable by studying the redshift of galaxies so far away that we view them as they were in an earlier cosmic epoch. If galaxies themselves looked different in their youth, that too should be observable. As Hubble optimistically forecast, "The combined efforts of mathematical physicists and observers have reduced the array of possible universes to such a limited range that it is now possible to predict with confidence that the type of the actual universe will be identified within the foreseeable future." To have even a chance to make that identification was, for young astronomers, a prize.

The generation that had led human perception out among the galaxies was coming to an end. The great opticians who built the California telescopes were dead— George Ritchey, the irascible perfectionist who called himself "commanding officer" of Mt. Wilson, but whose beautiful photographs helped open the world's eyes to the universe; and Bernhard Schmidt, hard as glass, who worked with an unlit cigar clamped between his teeth, glaring at the mirror on the grinding wheel as if it were a letter bringing bad news, when instead it was the heart of one of his remarkable Schmidt telescopes that made possible wide-field surveys of deep space. The astronomers were growing old. Baade accepted a university post in Germany where he could relax and stay out of drafty observatory domes. Humason, who remembered waiting on tables at Mt. Wilson, stayed on a few years longer, studying plates crammed with more galaxies than stars. Shapley retired from the directorship of Harvard College Observatory in 1952. Hubble died the same year. Immobilized by a heart attack during the final months of his life, he had observed by proxy, dispatching graduate students up the mountain with assignments precise and forceful as battle orders. The brightest of these students, in Hubble's view, was Allan Sandage.

A midwesterner like Hubble and Shapley, Sandage came from Iowa City, graduated in physics from the University of Illinois, and arrived at Cal Tech in 1948 when the graduate astronomy department there opened its doors. Knowing those doors would lead to the Mt. Wilson and Mt. Palomar telescopes, the program's organizers did not open them very far; Sandage and a little band of fellow sufferers flailed through what they all remembered as "incredibly hard" courses that seemed to leave no time for sleep. The reward for survival was a chance to complete the work of Shapley, Baade and Hubble, and perhaps answer some of the most important questions the twentieth century knew how to ask.

Knowing the depth of commitment the task would demand, Hubble went out of his way to impress his student with a sense that he was passing the torch. Sandage, a self-described "hick from the midwest," was invited to dinners at Hubble's home with guests of the stature of Aldous Huxley and Igor Stravinsky. "The conversation was intellectual, witty, very Cambridge high table," Sandage recalled. "There was lots of name dropping." If Hubble intended to dazzle, he did, and twenty years later Sandage still regarded him with awe. "A noble man," he called Hubble. "He interacted with people much as I imagine a god might."

Sandage's devotion to carrying on Hubble's life work led him to an isolation much like Hubble's. In part this resulted from envy of the publicity he received, when Sandage became the man the newspaper reporters called.

In part it was his almost religious intensity, his refusal to act like one of the boys, his insistence that what he was doing was every bit as important as it appeared, a quality that earned him the nickname "Super-Hubble." And in part it resulted from specialization; Sandage became the only person on Earth fully versed in observational cosmology. "Much of what Sandage is doing," said his colleague Jesse Greenstein, "he has been doing for so long that for anybody else just to catch up would take years. And nobody would consider retracing his work anyway, because he is viewed as a man of absolute integrity. I don't know any other field in the world where you can say that of somebody, that he has absolute integrity."

"People have attacked me because I do only one thing," Sandage said in 1975, "but that one thing is to try to figure out how the world is put together. The world is incredible—just the fact that you and I are here, that the atoms of our bodies were once part of stars. They say I'm on some sort of a religious quest, looking for God, but God *is* the way it's put together. God is Newton's and Einstein's laws."

A moment later Sandage laughed. "Anyway, I'm a nut, you know. Crazy."

5

THE CREATION OF THE UNIVERSE

Seek simplicity, and distrust it.

—Alfred North Whitehead

Sir Arthur Stanley Eddington, who introduced the astronomers and cosmologists to each other, was a Quaker, a mystic, a geometer, philosopher, astrophysicist, popularizer of science, and perhaps the greatest astronomer of his age, though it is difficult to say just what age that was. His popular essays on science sound modern today, while his philosophy of science—and he frequently mixed the two, page and paragraph—was widely viewed as a retreat to the eighteenth century.

He was born December 28, 1882, at Kendal, Westmorland. Numbers were his first preoccupation, almost his god. At age four he tried to count the stars in the sky. He mastered the multiplication tables through 24 before he learned to read, and when he *could* read, set out to count the words in the Bible. His widowed mother saved enough of her small income to send him to good schools. In the distilled competition of Cambridge mathematics, he won every prize.

As a scientist he helped diagnose how a star delivers energy from its crucible interior to the surface. He explored the non-euclidean geometries and produced a re-formulation of general relativity that delighted Einstein and introduced the theory to the English-speaking public. In his last twenty years he tried to unite quantum physics and relativity with what he called his own "mysticism," his conviction that the universe worth studying is the one within us. "It seems to me that the first step in a broader revelation to man must be the awakening of image-building in connection with the higher faculties of his nature," he wrote, "so that these are no longer blind alleys but open out into a spiritual world—a world partly of illusion, no doubt, but in which he lives no less than in the world, also of illusion, revealed by the senses."

A tall, lean, reticent man, Eddington seems to have been likable, if shy, and a rich conversationalist despite a habit of staring off into the distance; he was literally farsighted. He stayed at Cambridge most of his life, relaxing by reading detective stories and by playing golf. On the golf course he hit a slice and muttered to his partner, W. C. Fields style, "Space seems to be highly curved in this region."

In 1930, speaking to the Royal Astronomical Society, Eddington cited Hubble's new paper on the redshift-distance relationship as evidence that we live in an expanding universe. Eddington and a doctoral candidate, G. C. McVittie, had unearthed Lemaître's forgotten paper in the course of a research project, and now Hubble's work lent it the look of prophecy.* With great self-assurance, Eddington insisted that the universe really was expanding as Lemaître had proposed, and he held to this view even when Hubble, who found it melodramatic, withheld his support. At one point Hubble believed some of his own observations discounted the expanding-universe theory. Eddington coolly dismissed Hubble's worries, and further observations proved Eddington right.

Lemaître, rescued from obscurity, acclaimed as the man who predicted the expansion of the universe, resumed work on theoretical cosmology.

His immediate problem was to find a plausible explana-

* Friedmann too had predicted an expanding universe, but at this point nobody, including Lemaître, appears to have known of Friedmann's papers.

tion of how the expansion started. In his 1927 model, the one Eddington resurrected, Lemaître imagined the universe resting in a static "Einstein state" for an indefinite period before it began expanding. But this static cosmos could be made to seem mathematically reasonable only by using the "cosmological constant," and Friedmann had demonstrated that even with the constant included, a relativistic universe might expand. Hubble's results suggested the constant was unnecessary, and Einstein himself dropped it in 1931, publishing jointly with De Sitter the following year a simple expanding-universe cosmology that did not use it. Lacking the cosmological constant, Lemaître had to discard his original account of genesis. In the search for a new one, he turned to nuclear physics.

Lemaître proposed that the universe was born as a "primeval atom"—perhaps as little as 200 million miles in diameter—that disintegrated, sending matter flying apart. "Naturally, too much importance must not be attached to this description of the primeval atom," Lemaître wrote. It "will have to be modified, perhaps, when our knowledge of atomic nuclei is more perfect." The important thing was that the universe began violently, in "fireworks," as Lemaître put it.

Eddington, for all his affection for Lemaître, found this new idea distasteful. As an astronomer he was familiar with a universe of steady stars, cold space and galaxies perfect as flowers. The idea that everything had once been crammed into a hellish ball offended him. "Since I cannot avoid introducing this question of a beginning," he wrote, "it has seemed to me that the most satisfactory theory would be one which made the beginning *not too unaesthetically abrupt.*" (Eddington's italics.)

Eddington believed the laws of nature reside within our minds, are created not by the cosmos but by our perceptions of it, so that a visitor from another planet could deduce all our science simply by analyzing how our brains are wired. In Eddington's view, we know physical laws *a priori*, as Kant maintained, although where Kant conceived part of our *a priori* knowledge as inborn, Eddington felt it was derived from experience in observation and reasoning.

Most of us suspect that the world we see is in part genuine and in part distorted, or concocted, by our minds; the question is where the fulcrum stands between internal and external. Eddington put it farther toward the mind than did any other modern scientist. "We have found a strange footprint on the shores of the unknown," he wrote. "We have devised profound theories, one after another, to account for its origin. At last, we have succeeded in reconstructing the creature that made the footprint. And lo! it is our own."

The webs of knowledge spread outward and from time to time link strands, as they did for cosmology and astronomy in the 1930s. While a scientist may be patient about this process, expecting few final answers within his lifetime, a mystic, believing the answer is harbored within himself, is less inclined to wait. All he needs, he feels, is to find the key. Eddington looked for truth in the interplay of six apparently irreducible constants of physics—the velocity of light, the constant of gravitation, the charge of the electron, the mass of the electron and of the proton, and Planck's constant, which is a fundamental quantity in modern physics. The key for Eddington was N, possibly the largest number ever conjured up with serious intent by anybody. It stood for the total number of particles in the universe. Eddington estimated it at 10^{79}, or 10 million billion billion billion billion billion billion billion billion. He dropped it into the company of the physical constants, and things started to happen. The radius of the universe in centimeters, R (from Hubble's dwarf figures of the day), divided by the square root of N yielded 10^{-13}, which is close to 10^{-12}, the approximate radius of an electron in centimeters. The square root of N, 10^{39}, was roughly equal to the ratio between the weak (gravitational) force and the powerful (electrical) force that bind atoms together, calculated to be 2.3×10^{39}. Eddington felt that here were clues to the relationship between the very large and the very small, the secret of how, as he put it, an electron knows how large it ought to be.

Another number that fascinated Eddington was 137. Known to spectroscopists as the fine-structure constant, it incorporates the charge of electrons, the speed of light, and Planck's constant. Eddington hung his hat on cloakroom pegs numbered 137. The significance of 137 for him resulted from a complicated system, virtually a reconstruction of relativity along new lines, that he evolved in an attempt to bring the cosmos at large into local physics equations. Eddington intended the system to incorporate not only the behavior of atoms but possible ways they *might* behave (called "degrees of freedom"), as well as the context . required to make epistemologically meaningful measurements of them, and even the question of whether they existed (this codified in a mathematical tug of war, with non-being represented by Minus One). The number of elements Eddington required to describe physical process at a given point in space and time, using this approach, was 136. Well, 136 is pretty close to 137, the fine-structure constant. Eddington attempted to account for the one-digit difference by invoking a term in nuclear physics called the packing fraction, but later investigators who took the trouble to retrace this step found it highly questionable.

Eddington may have been the most lucid writer who was also a cosmologist since Lucretius, but his attempt at unified science and philosophy, published posthumously in a book titled *Fundamental Theory*, was a jumbled landscape viewed by moonlight. It may never be fully understood. Scholarly interest ebbed when both Eddington's N, the number of particles in the universe, and R, its radius, were shifted to higher values by the observations of Baade and Sandage, with the result that many of the coincidences of nuclear numbers with astronomical numbers ceased to coincide. Also, the theory was almost impossible to test. As Poincaré had said, physical theories, "above all, must lead to predictions," and Eddington's did not. Today it is of interest primarily as a historical curiosity.

Yet Eddington was a giant. He shared with Mach and Einstein an understanding that the universe is of a piece. "I only want to make vivid the wide interrelatedness of things," he said. He was set upon the course of his last twenty years in part by contemplating Einstein's equation $E = mc^2$. The world may know this equation as the key to nuclear bombs, but to Eddington its significance lay in the tie it established between the tiny atom and the interstellar world of light rays. The equation "links the universe to the atom," he wrote. He wondered: Why should the nucleus of an atom know anything about the speed of light?

If civilization survives its discovery of $E = mc^2$ for long, physics may well grow toward Eddington's vision rather than away from it. Twenty years after his death in 1944,

F. P. Dickson of the University of South Wales wrote, "A fair verdict is not that Eddington was 'off the beam,' but that there was as yet no beam and he was trying to show where a beam might be. The attempt was premature but it was beautiful."

George Gamow shared Eddington's conviction that science could learn about the cosmos by studying atoms. But Gamow, an irreverent, broad-brush scientist so casual about detail that he often failed to get dates and addresses right, was quite a different man than Eddington. One of his first contacts with Eddington's work came as an attempted practical joke.

As early as the 1920s, less gifted scientists had taken to imitating Eddington's cosmic numerology. Well-meaning papers appeared calling attention to previously unnoticed coincidences between this number and that. More practical-minded researchers regarded these efforts as on a par with reading tea leaves or studying the entrails of sacrificed beasts. In 1931, three postdoctoral fellows at the Cavendish Laboratory, Oxford, wrote an arch paper titled "Concerning the Quantum Theory of the Absolute Zero of Temperature." So carefully worded and closely reasoned was this short note—with references to "degrees of freedom" being "frozen out" in the style of the Eddington disciples—that the editor of the prestigious *Naturwissenschaften* published it without realizing it was a hoax. A few leading theoretical physicists were taken in too. When the editor learned the truth he was furious. The note concluded, incidentally, that all crystal lattices could be interpreted as constructed according to Eddington's favorite number, 137.

Gamow, twenty-seven years old and an accomplished practical joker, "could not sleep for a week," recalled Max Delbrück, his roommate at the Institute of Theoretical Physics in Copenhagen. "Somebody had outdone him." With co-conspirators Wolfgang Pauli and Leon Rosenfeld, Gamow waited for a serious Eddingtonian paper to appear in the *Naturwissenschaften.* Soon one did, an earnest contribution titled "Origin of Cosmic Penetrating Radiation." Gamow and Rosenfeld immediately wrote the journal's editor from separate cities, expressing sympathy that he had again been hoaxed and sorrow that another retraction would now be required. The plot faltered only when Pauli took pity on the man and balked at adding his weight to the deception.

Born in Odessa, Gamow left the Soviet Union, where Stalin permitted neither relativity nor Mendelian genetics to be taught in the schools, in 1933. Delbrück described him as "very tall and thin, looking even taller for his erect carriage, blond, a huge skull, and a grating high-pitched voice." Skeptical about religion from an early age, Gamow as a boy had concealed a bit of communion wafer in his cheek, smuggled it home, examined it under a microscope and decided it was bread and not flesh. "I think this was the experiment which made me a scientist," he said. He liked to tell the story of how the Russian mathematician Aleksei Nikolaevich Krylov calculated the distance from Earth to the throne of God as nine light-years, because when Russian churches directed prayers against the enemy during the Russo-Japanese war in 1905, they had to wait eighteen years (nine years for their prayers, traveling at the speed of light, to reach God, and nine more for His wrath to be transmitted back) before the great earthquake of

1923 struck Japan. Gamow rode to scientific conferences around Europe on a motorcycle, loved hiking, sailing, sketching and jokes, loathed boredom. "George had a very acute sense of the value of time," wrote the physicist William Fowler. "I remember once at a very dull and boring meeting in New York he turned to me and said, 'Sonny Boy, why are we wasting our time here? Let's go have a drink.' Which we did."

Gamow developed a quantum theory of radioactive decay, created an important model of the atomic nucleus, contributed to astrophysics and to studies that led to decoding the DNA molecule, and authored the modern approach to Lemaître's expanding-universe cosmology, for which he coined the term "Big Bang" theory. His approach to these matters was unfettered by exactitude; he could neither spell nor do arithmetic with any consistency and he had no particular desire to improve, recalling fondly that a mathematics professor at Odessa, when a student pointed out that he had made a numerical error in his calculations, roared back, "It is not the job of mathematicians to do correct arithmetic. . . . It is the job of bank accountants." When Gamow read a paper that interested him he often wrote the author a postcard or a letter, full of minor errors but likely to incise to the heart of the issue. Baade got a Gamow postcard soon after announcing his discovery that stars belonged to two populations. "Please tell me where the lower branch of the color-magnitude diagram joins the main sequence, and I will tell you the age of your Population II stars," it read. Baade replied that not enough about stellar evolution was yet known to make the extrapolation. A second postcard arrived by return mail with a drawing of the diagram. "I have extrapolated the lower branch thus," Gamow wrote. "OK, four to five billion years." Subsequent research traced the curve in the shape Gamow had drawn it, though his figures and Baade's were later revised upward. "That Gamow hit it so well was an accident," Baade said, "but his remark really contained the whole story of the interpretation of the diagram."

Gamow had studied under Friedmann, the mathematician who suggested that the universe should be expanding. Gamow was twenty-six years old when Hubble discovered the redshift-distance relationship. In 1934, while Eddington was calling attention to the connection between Hubble's observations and Lemaître's theory, Gamow arrived at George Washington University, where he was to formulate a new version of Lemaître's cosmology. He had begun thinking of atoms as artifacts of creation.

Most atoms have long lives. Those that form our bodies were in the oceans, soil and atmosphere of Earth before life here began, and they will be here long after we are gone. When the universe was thought to be infinitely old, it was theorized that the interiors of stars, where old atoms are constantly broken down and new ones created, probably acted as the forge where the atoms of our environment were created. Early attempts to take a census of the elements in the universe gave support to this theory. Hydrogen, the lightest of the atoms, was found to be by far the most abundant in the cosmos. Helium, second lightest, is the second most abundant. The amount of energy bound up in an atom's nucleus is in a general way proportional to its mass, and so some physicists concluded that atoms were formed preferentially according to how much energy they harbored. There is more hydrogen around than heavy elements, they argued, because it is easier to make hydrogen. A sort of stellar laziness was thought to rule.

On further study this relationship fell apart. It was found that although the cosmic abundance curve does decline from the lighter toward the heavier elements, it flattens out dramatically less than halfway down the periodic table, at about the position of zinc. Many of the heavier elements exist in approximately equal quantities even though their masses differ widely. For example, lead has over twice the atomic weight of rubidium, but the cosmos appears to contain about the same amount of lead as rubidium.

Gamow felt that this situation was inconsistent with the theory that the elements were created inside stars. To his mind the cosmic abundance curve suggested the results of a gigantic explosion, like a thermonuclear bomb. Or the Big Bang.

Gamow began to think that if he could reconstruct the conditions under which the elements formed in their observed abundances, he might learn what the universe was like in the first moments after creation. "The relative abundances of various atomic species," he wrote, ". . . must represent the most ancient archeological document pertaining to the history of the universe." The question was how to read the document.

During the war Gamow was a part-time consultant to the Applied Physics Laboratory at Johns Hopkins. There he met a graduate student, Ralph Alpher, who had just abandoned the subject of his doctoral thesis when a physicist at another school beat him to the result. Casting around for a fresh area of study, Alpher came across a paper that aroused his curiosity. Donald Hughes of the Brookhaven National Laboratory had measured a quantity called the neutron capture cross-section for a variety of atoms, and found that this quantity increased sharply throughout the first half of the periodic table of the elements, then flattened out, like an inverted version of the cosmic abundance curve. Neutron capture therefore might have something to do with the way elements were formed in the Big Bang. It might serve as a way to understand the cosmic abundances.

The product of this line of thought, a brash paper attempting to reconstruct events in the fireball of creation, appeared in a 1948 issue of *Physical Review*—on April Fool's Day, to Gamow's delight. Hans Bethe at Cornell was surprised to find himself listed with Alpher and Gamow as an author of the paper, which he had not worked on; Gamow had added Bethe's name because he thought it would be nice to have a paper on genesis written by three whose names began with Alpha, Beta, and Gamma, the first three letters of the Greek alphabet. Bethe had been one of the Cavendish Laboratory students who wrote the Eddington satire seventeen years before, and Gamow may have felt that by including the unwitting Bethe he was finally matching that joke.

The Alpher-Bethe-Gamow theory, like Lemaître's, described the universe as beginning in a highly compressed state, but while Lemaître's "primeval atom" was made of densely packed matter, Gamow's was composed of almost pure energy with only trace elements of matter. Lemaître's model resembled nuclear fission, the basis for the atomic bomb; Gamow took his from nuclear fusion, the hydrogen bomb. "One should imagine the original state of matter as a very dense over-heated neutron gas," Gamow wrote. He called this stuff Ylem, after the ancient Greek term for the chaos out of which the world was born. The Ylem cooled rapidly in the moments following the Big Bang; after two seconds it was cool enough for elements to begin forming, a process which trailed off about sixteen hours later. At a point 10,000 to 100 million years or so ABE—After the Beginning of Expansion, as Gamow liked to call it—the universe had cooled sufficiently that its energy content dropped below its matter content. It has been dominated by matter ever since.

As a catalyst for provoking thought about creation, the theory was a success. In terms of its original goal, accounting for cosmic element abundances, it failed. With the help of two other Johns Hopkins physicists, Robert Herman and James Follin, Jr., and using a borrowed Bureau of Standards computer with an input board that resembled a bowl of spaghetti, Alpher tried to work out the enormously complicated neutron-capture profile required to take the model from Ylem to a universe of elements. Alpher, Herman and Follin found that they were blocked at atomic weight five and again at eight. The universe contains no stable atoms with those atomic weights, and those made artificially in laboratories quickly fall apart. Gamow and his collaborators could find no convincing way for the putative element factory of the Big Bang to have bridged the two gaps. Since the first gap, atomic weight five, falls just above helium, the researchers were unable, for all their trouble, to cook up any elements other than hydrogen and helium in the Big Bang. Gamow consoled himself that 99 percent of the atoms in the universe are hydrogen or helium, so that he had accounted for the creation of all but one percent of everything, but it was frail comfort. The mystery of the cosmic element abundance was unsolved.

Evidence now suggests that some heavy atoms are indeed made inside normal stars, while others, very heavy elements such as gold and uranium, form in supernovae and then are blasted into space. If so, the gold in our jewelry came from stars that exploded long ago. The idea that heavy elements below gold were cooked up within healthy stars finds support in the fact that Baade's old Population II stars are poor in metals, as would be expected if they were born before other stars had time to seed the galaxy with heavy atoms. How normal stars vent their cargo of heavy elements into space, though, if they do, is still not fully understood.

The cardinal legacy of Gamow's Big Bang theory was a detail scarcely noticed by its authors. If the universe was once very hot, and has been cooling ever since, it ought now to be cold—as it is—but not *absolutely* cold. The energy of the original flash should still be around, though greatly dissipated by the expansion of the universe. In 1948, Alpher and Herman, cleaning up Gamow's faulty arithmetic, twice pointed out in print that the temperature of the cosmos today ought to be about five degrees above absolute zero. The residual energy would take the form of a low-level background radiation coming from all directions at almost equal strength.

It seemed an incidental matter at the time. Few scientists had any idea how to listen for a ubiquitous background radiation; the observation would probably require rockets sent above the atmosphere or sophisticated radio gear. Gamow, Alpher, Herman and Follin themselves never seriously considered searching for it.

In 1956 Gamow left George Washington University for the University of Colorado, and the team that put together the Big Bang theory disbanded.

A decade later, when radio astronomy had developed to a stature equaling that of optical astronomy, the background radiation was found in one of the most remarkable accidents of science. A certain amount of wistful hindsight ensued. At the Hale Observatories, Allan Sandage recalled once talking briefly with one of the Gamow team—he wasn't sure which one, Follin perhaps—in the early 1950s, when they speculated briefly that a rocket might be sent up with instruments to listen for the background radiation. "Nothing came of it," Sandage said, "but for a while there, he had the creation of the universe right in the palm of his hand."

6

THE ECHO OF CREATION

What Is Man?
The Sun's Light when he unfolds it
Depends on the Organ that beholds it.
—WILLIAM BLAKE

THE COOL NIGHT SKY THAT PRESENTS ITSELF TO OUR VISION IS a guise. Starlight, all light, rich and beautiful to our gaze, makes up only a sliver of the electromagnetic spectrum through which nature displays itself. The cosmos radiates most of its energy in wavelengths we cannot see. An important excursion into these invisible zones of the spectrum came with investigation of wavelengths much longer than light, those we call radio.

Radio was broadcast artificially well before it was discovered in nature—the young German physicist Heinrich Hertz propagated radio waves in a laboratory in the 1880s, and the Italian Guglielmo Marconi developed a working transmitter at the turn of the century—and as a result the word "radio" retains a connotation of artificiality, like "automobile" or "aspirin." The universe, however, has been in the radio business from the outset, and paints its own portrait in those wavelengths as vigorously as in light. If our eyes were tuned to radio (for which purpose they would need be the size of serving platters) the sky would look quite strange—the stars faded except for a few flashing pulsars, the galaxies swollen to several times their optical dimensions, the Milky Way a loom of fire. The universe produces radio radiation in a variety of ways. Once astronomers began to learn about radio, it looked newly rich and strange.

When Hertz discovered radio waves, a few scientists, notably the physicist and spiritualist Sir Oliver Lodge, predicted that there could be such a thing as radio astronomy, but the suggestion ebbed for lack of the technology needed to try it out. The first radio waves to be heard from space were intercepted accidentally by an engineer who was listening for something else.

Karl Jansky, employed by the Bell Telephone Company at their research laboratories in Holmdel, New Jersey, was assigned in 1931 to look into causes of static on long-distance phone lines. In an open field Jansky built what was in effect a set of exposed telephone cables that could be aimed to determine where static was coming from. The antenna consisted of eight steel rods shaped like squared-off croquet hoops, 12 feet high and mounted in pairs, set on a spindly wooden frame. The assembly, 100 feet long, rested on four spoked airplane wheels and a circular brick foundation. Its receiving cables were wired to an amplifier and moving-paper recorder. Jansky could determine the direction of radio noise by pushing the antenna around in a circle and noting where the recorder hit a peak. A photograph made at the time shows Jansky, in knickers, adjusting the antenna, which looks something like a large unfinished biplane.

Within a few months Jansky had established that noise afflicting long-distance callers fell into three categories. Most came from lightning in nearby thunderstorms, as expected. Additional noise came from distant thunderstorms, and Jansky's finding that this level was higher than anticipated made the experiment a success from Bell's point of view. In addition, Jansky picked up a third noise he could not explain, a thin, persistent hiss that did not vary with changes in the weather, and seemed to come from one spot in the heavens.

At first Jansky thought it was radio from the sun, something Lodge had predicted in 1894, but as months passed and the sun's position against the background stars shifted, the position of the radio source remained synchronized with the stars, not the sun. By 1933 Jansky estimated that it was located at roughly the celestial coordinates 17 hours 30 minutes by minus 30 degrees. Consulting a star chart, he found that this coincided with the center of the Milky Way. Somewhat reluctantly, Jansky concluded that Earth was being flooded with radio waves emanating from the heart of our galaxy. It was the first of many lessons to be learned about the strangeness of the radio sky, where distant galactic powerhouses outshine nearby stars.

Jansky proposed that someone build an antenna intended for astronomy. His antenna worked best at the relatively long wavelength of 14 meters, where resolution, the ability to focus clearly, is inherently low. An antenna built to scan the skies at wavelengths of one or two meters, he suggested, would resolve cosmic radio sources clearly enough to make possible genuine radio mapping of the sky.

Few observatories knew of Jansky's work and none were prepared to finance such a project, but a radio engineer and amateur astronomer named Grote Reber was. In the back yard of his home in Wheaton, Illinois, a Chicago suburb not far from the site of George Ellery Hale's old childhood observatory, Reber bolted together the world's first genuine radio telescope. It looked remarkably like its modern descendants. Its 31-foot metal dish reflected radio waves to a receiver suspended at the focus by four outstretched framework arms. A double horseshoe mounting allowed the dish to be pointed in altitude, while the rotation of Earth took care of east-west scanning. The telescope sat behind Reber's house, drawing tourists.

While mapping the radio sky, Reber took a course at the University of Chicago to refresh his understanding of astronomy. Then he took his maps and charts to Yerkes Observatory and tried to persuade the astronomers there that he had something worth looking at. Otto Struve was the observatory's new director. He recalled that Reber "brought to Yerkes a large stack of tracings, on which his instrument had recorded the intensity of radiation from the sky at a wavelength of slightly less than two meters. A definite increase of intensity was shown by a general bulge of the red line, drawn by the recording instrument at the exact time the Milky Way passed over his antenna. The presence of numerous violent and sharp disturbances was an annoying factor, which Reber explained as the result of various electrical appliances, such as a dentist's drill a block or two away, a trolley line of a street in his vicinity, or a passing car's faulty ignition."

Struve realized that because radio waves pass through clouds of interstellar dust and gas that block light, the center of our galaxy, forever hidden from optical astronomers, could be observed directly in radio. Most other astronomers, however, were skeptical, even hostile, about this eerie sky that could be observed day and night, where bright stars were invisible, the sun was a dim blotch, and the core of the galaxy showed up like the bones of a patient under a fluoroscope. When Reber submitted a paper on his work to the *Astrophysical Journal* the referees rejected it as unbelievable. Fortunately Struve was editor and printed it anyway, over their objections. Baade, Rudolph Minkowski and Greenstein were among a handful of astronomers whose interest was aroused. "We were beginning to realize," Struve wrote, "that a completely new branch of astronomy was emerging."

The rapid development of radar during World War II, especially by the British, created the technology radio astronomy would later employ, and led to a few accidental discoveries as well. A civilian on wartime duty for the British army, J. S. Hey, set up a system for detecting radar jamming by the Germans, was puzzled by a flood of four-meter to eight-meter wavelength radiation received in February 1942, and traced it to a solar eruption centered on a large sunspot. That same year a Bell Laboratories researcher discovered thermal, or "quiet sun" radio radiation. A radar array built on the English Channel to detect V2 rockets picked up the Milky Way background Jansky and Reber had studied. The scientists involved worked in virtual isolation, their work classified.

Astronomers in the Netherlands, prevented by the Nazi occupation from observing, turned to theoretical work. Jan Oort, who had helped analyze the structure of the Milky Way from photographs and had learned of Jansky's and Reber's discoveries, encouraged one of his students, H. C. van de Hulst, to consider the theoretical side of the matter and see whether he could determine how cosmic radio waves were being generated. Since there is more hydrogen in space than anything else, Van de Hulst concentrated on studying the hydrogen atom. In 1944, just as the occupation was ending, he announced that hydrogen atoms in space theoretically ought to emit radio radiation at a wavelength of 21 centimeters. I. S. Shklovsky in the Soviet Union independently made the same prediction.

Van de Hulst and Shklovsky's work drew upon studies of the magnetic properties of atoms made by nuclear physicists including Bethe, Niels Bohr, and Enrico Fermi. At issue was a slight magnetic instability in the hydrogen atom that causes each atom to disgorge a quantum of energy from time to time. This is a very rare event—it befalls a given atom only once in about 11 million years— and interstellar space contains only about one hydrogen atom per cubic centimeter, less than is found in the most perfect vacuum attainable in laboratories on Earth. But there is so much space that the total number of hydrogen atoms, each letting out a single *peep* of energy once every 11 million years, amounts to a continuous radio roar. Calculations indicated the radio radiation should have a wavelength of 21 centimeters. Galaxies are loaded with hydrogen and so presumably could be mapped in detail at the 21-centimeter wavelength. The rate galaxies were rotating also could be ascertained by measuring Doppler shifts of the 21-centimeter line. As the war drew to a close, a race began to detect this important form of radio radiation in the sky.

At Leiden, Oort and a colleague, C. A. Muller, built a 21-centimeter receiver and hooked it up to a 24-foot dish antenna owned by the Dutch Post and Telegraph Service. But their hastily assembled receiver caught fire, and before they could repair it, the discovery was made by two Harvard men, Edward Purcell and E. M. "Doc" Ewen.

Purcell, who was about to win a Nobel Prize for his studies of nuclear magnetic resonance, had worked on radar during the war. He had the expertise to design a receiver to detect the 21-centimeter signature of hydrogen atoms. When Ewen, a graduate student interested in microwave radio and radar, told him about Van de Hulst's theory, Purcell said, "Hell, we don't need to know that much astronomy, let's see if we can do it."

Their antenna was a wooden horn the size of a bathtub, lined with copper foil and stuck out a window of Harvard's physics laboratory. It cost $400. Passing students threw snowballs into it. Most of the electronic equipment was borrowed. Twin receivers were used, sweeping through the 21-centimeter frequency one after the other, so that spurious radio noise, showing up on both simultaneously, could be ignored, while true hydrogen radiation would reveal itself as a peak when each receiver in turn crossed the 21-centimeter wavelength. On March 25, 1951, Purcell and Ewen recorded 21-centimeter radio. They noted that it reached a maximum when the horn was pointed near the center of the Milky Way.

Six weeks later, their charred electronics repaired, Oort and Muller duplicated the observation, and an Australian team did the same in July.

Radio astronomy developed rapidly in the years that followed. Large radio telescopes were constructed in Australia, the Soviet Union, the United States, and in England, where they were particularly welcome to British astronomers long hampered by cloudy skies. The discoveries of the new branch of science included exotica like the pulsars—rapidly spinning crushed stars left behind after a supernova—and radio galaxies, blasting out too much energy for theorists readily to explain. It was a period when novelty eclipsed retrospection. Few remembered the theories of Gamow, Alpher and Herman in any detail.

We live in an ocean of whispers left over from our eruptive creation, Gamow had said. Nobody was listening.

Gamow was interested in the Big Bang as a way the elements might have been created, but his theory was revived nearly twenty years later by investigators interested not in creating elements but in destroying them. Robert

Dicke and his Princeton colleagues P. J. E. Peebles, P. G. Roll and D. T. Wilkinson were looking into the possibility that the universe oscillates, expanding for billions of years, then collapsing into a fireball that might, in turn, erupt to create a new cosmos. Dicke was concerned with whether the heat generated in a collapsing universe would be sufficient to strip down all the atoms, so that each new universe could start fresh without any trace of its predecessor. If so, then any question of reading cosmic history prior to the Big Bang would be expelled from scientific concern, since all information from the prior epoch would have been obliterated. If not, studying the traces that survived the Big Bang might tell us something about the universe that preceded it. Dicke and his colleagues estimated that a fireball of one billion degrees absolute—a good "hot bounce," as they called it—would have been required to wipe the slate clean. This avenue led them, from an opposite direction, to the same sort of calculations Gamow had made.

They proceeded with their speculations in 1964 under the pall of what Dicke later admitted was "remarkable ignorance" of Gamow's studies. As they detailed the theoretical conditions of a "hot" Big Bang, it dawned on them, as it had on Gamow, Alpher and Herman in the 1940s, that energy from the explosion, an echo, so to speak, should pervade the universe today.

Dicke had worked in radio with Purcell during the war, developing the Dicke radiometer, a sophisticated instrument for detecting microwave (that is, very short wavelength) radio radiation. In 1946 he had pointed a radiometer at the sky and determined that any short wavelength cosmic radiation (which he assumed would come from galaxies) must lie at energies below 20 degrees Absolute.* A paper to this effect, written by Dicke and three associates, appeared in the *Physical Review* shortly before Gamow's Big Bang theory was published in the same journal; one may take out the volume today and read Gamow's team predicting a 5-degree background radiation while, a few dozen pages away, Dicke reports that microwave cosmic radiation must lie below 20 degrees. But the two groups never made contact. Dicke did not have the Big Bang in mind in those days; Gamow did not know an instrument existed of a sort that could detect its echo.

When Dicke resurrected the idea of a cosmic background radiation in 1964, Roll and Wilkinson set to work building a modern Dicke radiometer to listen for it. The device was small, resembling an old acoustic phonograph, but elegant as a piece of jewelry, with liquid helium to stabilize the circuitry and act as a reference for checking cosmic temperature, and a gold-plated horn pointing skyward. Meanwhile Peebles began writing a paper describing what they expected to find. The background radiation, if it really originated in the Big Bang, should follow a "blackbody" curve—an idealized trace of energy against wavelength predicted by quantum theory. Its temperature would depend upon the total energy of the Big Bang. Peebles was confident the radiation should still be powerful enough today, some 15 billion years or more after the fact, to be observable in microwave radio.

That same year, two Soviet astronomers named Doroshkevich and Novikov were reviewing the old Gamow papers and came across the background radiation prediction.

---

* In the microwave region it is convenient to express signal power as temperature.

They realized the radiation might be observable by radio. They urged that the microwave region was a promising place to look for it, and even proposed which of the world's radio antennae was best suited for the job—a horn antenna operated by Bell Laboratories in Holmdel, New Jersey, thirty miles from where, unknown to the Soviets, Dicke and his coworkers were laboring on the same question.

Long radio waves, like big fish, can be caught in loose nets, but microwaves, like little fish, require a tighter mesh, and so the Bell horn was built solid. It looked like an alpenhorn the size of a boxcar. A scoop at one end admitted the microwaves. The horn funneled them to a focus in a clapboard shed built on stilts. In the portion of the electromagnetic spectrum for which it was built, the Holmdel horn ranked among the most sensitive radio telescopes in existence, but it had been used primarily for satellite communications, not astronomy. The Dicke group had never given it much thought.

Arno Penzias, a young researcher from Columbia, was lured to Bell in 1961 with a promise that if he applied his talents to perfecting the horn as a communication device, he would later be allowed to use it for astronomy. He was joined the following year by Robert Wilson, a Rice graduate who did his graduate thesis on radio astronomy at Cal Tech.

The two made a good team. Penzias was intellectually aggressive, a generalist interested in music and literature as well as science, conceptually bold if sometimes offhand about specifics. Wilson was less talkative, more precise, a demon for accuracy; he had the patience to wrestle nominal performance from temperamental, state-of-the-art electronics gear. Penzias was given to enthusiastic monologues, drawing grand interconnections among ideas, while Wilson listened with a smile, interrupting occasionally to correct a figure or define a term.

Bell, with considerable stake in the embryonic communications satellite business, ordered the microwave horn modified so it could talk to the first Telstar satellite. Plans called for Telstar to relay telephone conversations and television signals across the Atlantic by using a Bell antenna in Andover, Maine, on one end and a new antenna in France on the other. But as the Telstar launch date approached, the French antenna was still under construction, and Bell executives began to worry that they might find themselves with an expensive satellite in orbit and only a single facility on the ground ready to communicate with it. "It would be kind of poor to transmit to a satellite and then receive from the satellite in the same place," Penzias said. "You could do the experiment using only the Andover antenna, but there would be something psychologically wrong with it. Somehow, you couldn't bring yourself to gather the newspaper reporters together, send out a pulse, and tell them, 'Well, the signal has been up there and come back.' You want the signal to arrive somewhere else. They weren't sure France would be ready, and wanted another receiver prepared just in case."

To do this, the important thing was low noise. The less spurious noise in the receiver, the clearer the satellite pictures would be, and the more promising Bell's debut in the field. To attain the best possible signal-to-noise ratio, Penzias and Wilson fitted the Holmdel horn with a maser (the word is an acronym for Microwave Amplification by Stimulated Emission of Radiation), a device that stored energy and released it when "tickled" by a weak signal of the proper frequency. At the heart of the maser was a ruby crystal embedded with chromium atoms that had to be chilled to nearly absolute zero. Penzias and Wilson jammed four-foot-high yellow steel bottles of liquid helium, temperature 4 degrees Absolute, into the receiver shed and hooked them up by flexible tubes to a jacket surrounding the maser. Valves froze and burst, filling the cramped enclosure with clouds of helium, but when the plumbing behaved the maser worked fine.

As it happened, the French got their antenna into operation by the time Telstar was launched and signals were beamed across the Atlantic using all three stations, bringing the project to a successful conclusion. Penzias and Wilson now wanted to get on with radio astronomy, using the horn and the new maser. First Bell had one more assignment. Company engineers needed a precise measurement of the horn's sensitivity, so that they could be sure Telstar had been performing up to specifications when it relayed signals back to Earth. The engineer's term for the figure needed was "gain," meaning just how much of the measured signal was "gained" from Telstar, as opposed to the amount it was boosted by the receiver's amplifier circuitry. To get the answer, Penzias and Wilson sent a helicopter flying around Holmdel with a small microwave transmitter aboard, aimed the horn at it and recorded the results. Knowing the exact distance of the helicopter and the precise power of its transmitter, they could deduce the gain of the antenna.

The job was mostly drudgery. Penzias and Wilson consoled themselves that if they could establish gain to a high degree of accuracy, they could then go ahead to make some scientifically worthwhile observations of the absolute intensity of astronomical radio sources. This would be helpful to other radio astronomers, who seldom knew the exact strength of cosmic radio objects they observed because they rarely took time to measure the gain of their antennas. So, it was hoped, a little original science might be rescued from an otherwise tedious exercise.

When Penzias and Wilson reduced their data, the results were good enough for the Telstar project but not for astronomy.

The problem was that a low, steady, mysterious noise persisted in the receiver. It showed up on the chart recorders day and night, regardless of where in the sky the antenna was pointed. Such noise had been noticed before and had posed a problem ever since the maser was installed. It had prompted harsh words between the maser designers, who blamed it on the antenna, and the antenna designers, who blamed it on the maser. Now it had become an embarrassment to Penzias and Wilson. They resolved to identify its source if outside the antenna or clean it up if it were internal; to "exonerate the antenna" or fix it.

They took apart the maser and found it scrupulous. They found two pigeons nested inside the horn. A little of a pigeon's body heat might register in microwave radio. They sent the pigeons by company mail to Bell's offices in Whippany, New Jersey, sixty miles away. The pigeons turned out to be homing pigeons and were back inside the horn two days later. Penzias and Wilson removed them again. The temperature of the pigeonless horn dropped only half a degree. They took apart the narrow throat of the horn, replaced a few suspicious parts, scrubbed and restored it. This reduced the ceaseless noise only a tenth of a degree. Wilson crawled over every inch inside the horn, masking rivets with aluminum tape. The hiss remained. He and Penzias considered and rejected every natural or man-made explanation they could think of. Months passed. The hiss continued from every quarter of the sky.

"Now here we had a result we knew couldn't be right,"

Penzias said. "You could dispose of any radio source known to exist, if, for no other reason, because all radio sources known at that time radiated more at longer wavelengths than at shorter. Yet here we had this thing in short wavelengths. The only exception is blackbody radiation. And we knew—or thought we knew—there wasn't any blackbody radiation out there, because it was empty space. So we were stuck with a mystery."

So in 1964 three teams of scientists were working in a perfect tripod of ignorance. The Soviets had predicted that if Gamow were right and the universe began in a Big Bang, background radiation following a blackbody curve ought to pervade the sky and be audible to the Bell horn. At Bell, Penzias and Wilson were picking up just such a sound but did not know what it was. At Princeton, a half-hour's drive from Holmdel, Dicke and his team were preparing to search for the noise without knowing about Gamow, the Soviets, or Bell. And Gamow knew nothing of any of their efforts.

Penzias and Wilson prepared to hide news of what they regarded as their failure to account for the noise in the midst of a dry paper relating how they had calibrated the Holmdel antenna. "The early history of radio astronomy was full of incorrect results," Penzias explained. "This is true of any science when it first starts up, but we radio astronomers were terrified of making more mistakes. So we buried it in another paper. Figure the paper went on for twenty pages or so. In the middle of it you mention this residual noise problem. If you're wrong about it, if you overlooked some obvious cause, people won't point to it as the cardinal point in the paper. You're not sticking your neck out so far."

In December 1964, on an airplane returning from an

astronomical conference in Montreal, Penzias sat next to Bernard Burke of the Carnegie Institution, in Washington, D.C., and mentioned his troubles with the residual hiss. Not long afterward, back at Holmdel, he had a call from Burke asking how things were going.

"We still have this damn noise," he told Burke. "We're pretty sure it really exists, and it seems too important not to publish, but we're sure somebody will shoot us down for putting it out, so we're going to ditch it in this other paper."

Burke said he had just seen a preprint of the paper Peebles had been working on at Princeton, predicting that a background radiation of roughly 10 degrees intensity, left over from the creation of the universe, could be observed with microwave radio telescopes. Penzias might want to read it.

Penzias called Dicke, who sent him a copy of the still-unpublished Peebles paper. Penzias read it and called Dicke again. Dicke drove over and had a look at the horn and the results. He, Penzias and Wilson stared at one another.

To avoid potential conflict, they decided to publish their results jointly. Two notes were rushed to the *Astrophysical Journal Letters.* In the first, Dicke and his associates outlined the importance of cosmic background radiation as substantiation of the Big Bang theory. Peebles' paper had still not appeared, but he was credited for his work. A second note, signed by Penzias and Wilson and titled, "A Measurement of Excess Antenna Temperature at 4080 Megacycles per Second," began:

> Measurements of the effective zenith noise temperature of the 20-foot horn-reflector antenna ... at the Crawford Hill Laboratory, Holmdel, New Jersey, at 4080 Mc/s have yielded

a value about 3.5°K higher than expected. This excess temperature is, within the limits of our observations, isotropic, unpolarized, and free from seasonal variations (July, 1964–April, 1965). A possible explanation for the observed excess noise temperature is the one given by Dicke, Peebles, Roll and Wilkinson (1965) in a companion letter in this issue.

Purcell at Harvard read this modest announcement. "It just may be," he said, "the most important thing anybody has ever seen."

Penzias and Wilson found the dimensions of their discovery difficult to absorb. "It never exactly hit us all in one day," Penzias said. Wilson was surprised that it made the front page of *The New York Times,* in a bylined story May 21, 1965. "Scientists at the Bell Telephone Laboratories have observed what a group at Princeton University believes may be remnants of an explosion that gave birth to the universe," the story began.

"I think reading Walter Sullivan brought it home," Wilson said after reading the *Times* account. "It seemed larger then. More important."

Further observations at other frequencies tended to confirm that the background radiation followed a blackbody curve—which looks something like a roller-coaster run, a slow incline followed by a rapid drop—as Gamow's Big Bang theory predicted. The Princeton group got their radiometer operating in six months and observed the noise at 3.2 centimeters wavelength. (The Holmdel antenna was tuned to 7.35 centimeters.) Its intensity fell on the curve just where theory predicted. Astronomers at Cambridge University made a third observation at 20.7 centimeters, and this too conformed to the expected curve.

These observations were limited to the slow-rising side of

the blackbody curve. The expected peak of the curve lay at a wavelength where interference by Earth's atmosphere prevented ground-based observation. Ultimately rockets and satellites were used to chart the background radiation near its peak, but prior to that, several astronomers found an elegant way to make essentially the same observation indirectly, by studying molecules in space. Later, they learned that their work had almost been duplicated in the 1940s, in another of the many coincidences and near-misses that characterized the history of this curious subject.

A variety of molecules have been found floating in clouds in space, among them water, carbon monoxide (which Penzias and Wilson helped detect), and even alcohol and formaldehyde. Just how the molecules got there is not fully understood, but presumably some were formed when two compatible atoms bumped into each other in an interstellar cloud and, like lonely countrymen vacationing where no one else speaks the language, stuck together. Cyanogen, composed of one carbon and one nitrogen atom, was among the first interstellar molecules to be detected. It was observed in the 1930s, its presence betrayed because it added faint absorption lines to the spectra of stars whose light passed through it before reaching Earth.

An interesting thing about cyanogen molecules in space was that many seemed to be "excited," as physicists say, meaning that their energy levels were boosted above normal. In 1940, W. S. Adams at Mt. Wilson obtained spectroscopic evidence of excited cyanogen molecules in space, and Andrew McKellar of the Dominion Astrophysical Observatory in Canada then deduced, from the proportion of excited molecules in the cloud, just how much energy was stimulating them. He concluded the cyanogen was being warmed to a temperature of 2.3 degrees above absolute zero. McKellar and several other astronomers assumed the energy involved came from starlight. Gamow's theory had not yet appeared, and when it did, no one considered that the interstellar cyanogen might be heated not by stars but by the residual noise of creation. Gamow's associate Herman knew of the cyanogen studies but kept them in a separate mental compartment from cosmology. "I never put the two together," he said. "You know, in retrospect everything becomes clear."

Cyanogen's secret persisted even after Penzias, Dicke and the others published their notes on the background radiation. Penzias later fumed at himself for not having thought of this simple way of watching the echo of creation operating in space. "I knew about the cyanogen work," he said. "In fact, when I was working on another interstellar molecule before I turned to the microwave background radiation, I even talked to George Field at Princeton about it—*talked about* the fact that this interstellar cyanogen was at two point something degrees! But I never connected the two at all. Later I felt so stupid."

Field was at Berkeley when he learned that the background radiation had been discovered, and he realized almost immediately that it might be what was heating up the cyanogen clouds in space. At least two other American astronomers, and Shklovsky in the Soviet Union, independently reached the same conclusion, with great benefit for studying the background radiation. Cyanogen is excitable by radiation at 2.6 millimeters, a wavelength that happens to lie close to the peak of the blackbody curve predicted by Big Bang theory. By measuring the degree of excitation of interstellar cyanogen, astronomers therefore could virtually dip a thermometer into space and sample the radiation.

This was soon done. Cyanogen excitation levels were measured in diverse regions of space, and in every case the levels indicated a true cosmic temperature of between two and three degrees. It was the most powerful confirmation the violent genesis cosmologies of Lemaître and Gamow enjoyed. Lemaître, seventy-two years old, heard of it in one of his last scientific conversations.

Gamow, Alpher and Herman, the men most vindicated by the discovery of the cosmic background radiation, greeted the news with mixed emotions. Gamow was astonished to find that the discoverers' papers did not credit his foresight. Dicke mentioned Gamow's attempt to account for synthesis of the elements in the Big Bang, an effort which had failed, but was silent on his prediction of the background radiation, which now had triumphed. Peebles also was silent. So were Penzias and Wilson. Gamow wrote an indignant letter in his round scrawl to Penzias, listing three occasions he had prophesied the background radiation in print. He had even included it (though at an inflated estimate of 50 degrees, his arithmetic having proven undependable as ever) in his book *The Creation of the Universe*, a best seller with nearly a quarter million copies in print. "You see," Gamow wrote, "the world did not start with almighty Dicke."

Now in his sixties, Gamow occupied a cottage in Boulder, Colorado, wrote popular books on science, answered his voluminous mail with notes, line drawings or poems, drank too much, told friends that each day he felt more strongly that nature was profoundly simple. The dust in space between the stars, he wrote, "has about the same constitution as the dust clouds raised by a bulldozer working on a mountain road." At an astrophysics conference he chaired a session on the microwave background radiation and joked about the way his forecast of the echo of creation had been overlooked: "If I lose a nickel, and someone finds a nickel, I can't prove it's my nickel. Still, I lost a nickel just where they found one." In the last weeks of his life he had recurrent dreams of talking with Newton and Einstein. He died August 20, 1968.

His younger collaborators, Alpher and Herman, had dropped out of cosmology. Herman went to General Motors, where he applied his skills, sharpened in reconstructing the birth of the universe, to examining automobile traffic. He first developed a theory to predict the behavior of cars on single-lane highways, then moved to multilane traffic, then to crossings and complicated patterns like the entrance to the George Washington Bridge.

"Alpher and I get so damn bitter about this," Herman said. "We did all that work, but strange as it may seem, we didn't ever think that, by God, we should beat the drums for it, get some publicity, have somebody take a look for this background radiation. The astronomical community back at that time didn't show much interest in what we had been doing. My impression was that they were not particularly interested. I don't know. We've done a great deal of work in other fields, but I think our greatest love and satisfaction came from astrophysics and cosmology. The mathematics wasn't complicated as those things go. Our main role was to imaginatively conceive these things, and they turned out not to be just figments of the imagination. We did some things that stood up.

"Frankly, Alpher and I have been extremely upset by the literature which chooses to ignore what we did. It's very bad. People publish papers and don't refer to the fact that we predicted this business. We're very disappointed."

Alpher took a job at General Electric's laboratory in Schenectady, New York. When the discovery of Penzias and Wilson was published, he said, "I was thoroughly and completely delighted, and at the same time repelled by the fact that there was an accompanying paper by Dicke and three others with no reference whatsoever to our predictions. Here was a restatement of what we had said seventeen years earlier, with no reference to our work."

In 1971 Alpher gave up science and moved to an administrative position at GE. "I left science in part because I was kind of disillusioned by the poor scholarship of so many scientists," he said. "Not only Dicke but others. It was discouraging. I would still like to get back into physics at some time in the future, though, if I can still do physics."

7

AN ETERNAL UNIVERSE

According to my opinion, the same force and vigor remain always in the world, and only pass from one part of matter to another, agreeably to the laws of nature, and the beautifully preestablished order. And I hold, that when God works miracles, he does not do it in order to supply the wants of nature, but those of grace.

—Gottfried Wilhelm von Leibniz

In 1948, THE YEAR GAMOW PUBLISHED HIS BIG BANG THEORY, three British astronomers, of whom Fred Hoyle would become the most widely known, created the Steady State theory. This unique cosmology combined skepticism about the Big Bang with an attempt to restore the ancient concept that the universe has forever existed and will exist forever, infinite in time and space. As a theory it found little observational support and expired within twenty years. Its importance lay in part in the fact that it became a rallying point for those skeptical about the Big Bang theory in particular, and the grander claims of science in general.

Thomas Gold, Herman Bondi and Hoyle, all from Cambridge University, were assigned during the war to help develop admiralty radar, at Dunstable, a Canadian Air Force base northwest of London. Bondi and Gold rented a cottage near the aerodrome. Rent was low because the cottage sat next to a bomb dump. Canadian attack bombers returned from dawn raids along the French coast carrying live time-fused bombs they had been unable to drop because of flak damage or failure of the release mechanism. The bombs were made nondefusable. Anxious crews carted the ticking bombs to the dump and left them to explode. When they did they blew out windows in the cottage. The dull thud of the bombs going off alerted the local glazier, who selected glass panes of the correct size, rode out to the cottage with them on his bicycle, and replaced the broken windows. He billed the Canadian Air Force on a monthly basis. In this house two or three nights a week, Bondi, Gold and Hoyle met and discussed the nature of the universe.

To the extent that the three could be assigned roles, Bondi, a brilliant mathematician, handled the analytical end of their talks, while Gold, naturally skeptical, questioned assumptions, and Hoyle was the instigator. Tough-minded and unpolished, Hoyle grew up in a textile valley in the north of England and climbed to Cambridge on a ladder of scholarships. He retained a distrust of what he felt were "establishment" scientific ideas, among them the Big Bang theory. "Hoyle was the driving force, always asking questions of us," Gold recalled. "How could Hubble's observations be understood? He was full of Hubble in those days." Unsatisfied with the answers Lemaître and Gamow favored, and prompted by the time-scale problem Hubble had encountered, Hoyle groped for some new way of interpreting the expansion of the universe.

After the war the three kept up their talks at Cambridge, usually in Bondi's rooms at Trinity College. One evening Gold fell into a short monologue and heard himself asking, "Do we really know the universe must have come from a beginning? After all, we had to create the matter then, at the beginning. Why not create it a little at a time?" Suppose matter were being created continually in space, Gold said. Then the universe could expand yet never thin out. New galaxies, formed out of new matter, would fill the void between the old.

Gold later thought the idea might have been planted in his mind at the local movie theater when he saw *Dead of Night,* a film he remembered as beginning and ending with the same scene, lending it a circular, unending quality. Whatever generated the idea, Gold felt he had hit on something. Hoyle and Bondi were, at first, less impressed.

Bondi offered to explore its mathematical ramifications, expecting to find contradictions. Within three weeks he reported that instead, it worked rather nicely. Continuous creation, it was obvious from the outset, violated the law of conservation of energy, a foundation stone of physical science that states that the total amount of matter and energy in any closed system (including a closed universe) never changes. But so, it could be argued, did any of the fireball cosmologies that talked of "creation" in an instant long ago. And Bondi found that the extent to which continuous creation theory violated the law was not so great as it had first appeared. While new atoms theoretically appeared out of nothing, old galaxies meanwhile receded to near the edge of the observable universe, their light so redshifted it could hardly be detected. So, the matter-energy content of the *observable* universe might remain constant even though new matter was being created out of nothing.

Hoyle proposed that they work up a continuous creation, steady state cosmology and publish it jointly. But as the three talked they found themselves divided over whether to concentrate on the continuous creation aspect of the theory, as Hoyle preferred, or upon the Steady State aspect, favored by Bondi and Gold. The result was that Hoyle wrote one paper, Bondi and Gold another. Hoyle finished and submitted his first, to the irritation of Gold, who had been talking about the theory as his own idea. But the journals Hoyle first sent his paper to rejected it, so Bondi and Gold won the little derby. Their published paper was dated July 13, 1948, Hoyle's August 3.

The Bondi-Gold paper was philosophical in tone, light on mathematics and strongly influenced by the work of Edward Milne, an Oxford mathematician who had turned to cosmology in the early 1930s. Milne, always a mathematician at heart, esteemed deduction. The inductive method prevalent in science, with its emphasis upon observation and experiment, struck him as primitive. He hoped science would get over it. "The more advanced a branch of science, the more it relies on inference and the fewer the independent appeals to experience it contains," he wrote. Milne developed a theory of kinematic (that is, based upon motion) relativity, intending to replace Einstein's formulation with one deduced entirely from the passage of time. The theory was not well received, but elements of it enriched cosmological discussion for years. Particularly important to Bondi and Gold was what Milne called the "cosmological principle."

The cosmological principle held that models of the universe must allow everyone in the universe the same general view. There may be no privileged observers. Cosmology was not to repeat the pre-Copernican mistake of placing men in the center of things or otherwise attempting to account for the appearance of things by special pleading. The principle did not say the universe must look *exactly* the same from all perspectives; localities naturally vary. Observers near the core of spiral galaxies, if there are any observers there, bask in skies blazing with stars, while others, circling stars expelled from their parent systems, journey in the blackness of intergalactic space; large spirals like ours afford their residents a bounteous milky way, while others, the small irregular galaxies, are paltry as spilled salt. Milne's point was, rather, that the large-scale look of things from every point in the cosmos must resemble ours, that our perspective must be assumed ordinary, in any plausible model of the cosmos. As such, it has been more or less a rule of the game ever since.

Bondi and Gold took Milne's principle a step further. Since relativity requires that we regard the galaxies as inhabiting a continuum of time as well as space, they asked, why confine the cosmological principle to space alone? Why not demand that the universe present the same aspect to observers not just in all places, but at all *times*? Their answer was to postulate what they called the "perfect cosmological principle." Our overall outlook on the cosmos, they maintained, ought to be shared by observers in all quarters of the universe and at all times in its history. Evolving universe theories violate the perfect cosmological principle because they imagine the universe was once very different, before the galaxies formed, and imply that in the future, whether the universe expands forever or eventually collapses, things will be much different than at present. Only the Steady State theory, where new galaxies constantly form to fill unfolding space between the old, satisfies the principle its authors devised.

Hoyle preferred a less immodest approach, concentrating on the continuous creation facet of the theory. In his original paper, as well as in later versions he developed with J. V. Narlikar and in a revision by William McCrea, Hoyle's theory added a "creation" term to Einstein's general relativity equations, designated by the letter C. In such a theoretical universe, a feedback mechanism balanced the rate of expansion and that of creation of new matter. In part because his theory contained at least a bow in the direction of relativity, and therefore incorporated a theory of gravitation, as Bondi and Gold's did not, Hoyle became the leading advocate of the Steady State theory among scientists. His gift for popularization—he wrote books on astronomy and science-fiction tales—gained him the same role in the public eye.

As a wholly new cosmology the Steady State had its points. It described a universe of regal placidity, dressed in robes of the infinite. Lacking a fireball genesis, it called upon the stars to produce the heavy elements, and in that won out against Gamow, whose attempts to synthesize higher elements in the Big Bang failed.* There was, of course, no evidence that a single particle had ever appeared out of empty space, but this, curiously enough, was not a telling argument against the theory. The universe is mostly space, and the rate at which creation would have to proceed to counterbalance expansion is slow—only a couple of atoms must appear in a cubic city block per year to satisfy the theory. A physicist could scrutinize an empty liter jar for a million years and if not one atom appeared in it he would still not have much of a case against the theory.

As a philosophical question, which side to take was almost a matter of taste. If evolving universe cosmologists regarded the idea of atoms popping into being out of empty space as implausible, the Steady State advocates found the Big Bang equally so. The Big Bang, after all, requires us to accept that the cosmos was once crushed together at great heat and density, and we are permitted to inquire neither how large this universal furnace was, because no caliper could exist to measure it, nor how hot it was (Gamow's calculations deal with the first moments *after* expansion began), nor what it was doing before expansion commenced. Milne called the fireball a "grand irrationality." Hoyle admonished that "it is against the spirit

---

* Oddly enough, the chief legacy of the Steady State theory may be that Hoyle, striving to determine how stars form elements inside themselves, contributed greatly to astrophysics. One astronomer wrote, "Together with William Fowler of Cal Tech, Hoyle provided the essential insights required for the modern theory of stellar nucleosynthesis—and on such grounds alone Steady State cosmology has produced a major triumph in modern astronomy."

Globular clusters play the role of cosmic game preserves, where astronomers can study the oldest stars in our galaxy. About 100,000 stars belong to this globular, designated M13.

HALE OBSERVATORIES

Galaxies too congregate in clusters. In this photograph the bright objects with protruding defraction spikes, as well as the few dozen sharp points of light, are foreground stars. All the other objects are galaxies, members of the Corona Borealis cluster, one to two billion light-years distant.

HALE OBSERVATORIES

Jesse Greenstein, director of Cal Tech's department of astronomy, saw breakthroughs in understanding the evolution of stars, the structure of galaxies, and the scope of the cosmos.

CAL TECH

Walter Baade, a tireless observer, discerned a pattern in the evolution of stars that doubled estimates of the size of the universe.

HALE
OBSERVATORIES

Maarten Schmidt, puzzling over the mystery of the quasars one morning at his desk at Cal Tech, hit upon the interpretation that revealed them to be brilliant beacons near the outposts of space and time.

CAL TECH

Milton Humason, who arrived at Mt. Wilson as a mule driver, here measures redshifts in the light from remote galaxies, helping to determine the scale of the cosmos and the rate at which it is expanding.

Allan Sandage, the discoverer of quasars, discusses the expanding universe.

Light older than Earth is recorded on this plate of the extremely remote galaxy 3C295, observed by Hermann Minkowski in 1960, during his last night on the 200-inch telescope prior to retirement. Designated here by a white pointer, the galaxy looks almost starlike.

A spectrum of 3C295 obtained by Minkowski reveals that its spectral lines (center) are greatly displaced toward the red, compared to those of a laboratory reference spectrum (above and below). Interpreted in terms of astronomers' best estimates of the scale and dynamics of the universe, the redshift indicates that light from 3C295 traveled through space for seven billion years before reaching us.

The dawn of radio astronomy came when Karl Jansky accidentally detected, in 1931, cosmic radio waves. Jansky had built his antenna to investigate causes of noise on long-distance telephone lines.

In 1933, Jansky displays a chart of the natural radio noise he discovered coming from beyond Earth.

Grote Reber took up Jansky's work, compiling a radio map of the sky that helped convince astronomers radio was as natural a form of energy in the cosmos as light.

The first genuine radio telescope was built by Grote Reber in the back yard of his home in Wheaton, Illinois.

NATIONAL RADIO ASTRONOMY OBSERVATORY

Robert Wilson (left) and Arno Penzias (right), using the horn-shaped radio telescope in the background, inadvertently discovered the cosmic background radiation in 1965. Predicted twenty years earlier in George Gamow's Big Bang theory, the background radiation is believed to be residual "noise" from the violent event that initiated the expansion of the universe some 15 or 20 billion years ago.

BELL TELEPHONE LABORATORIES

Secrets of a quasar are probed in these two negative prints, favored by astronomers because they reveal greater detail than conventional white-on-black prints. At top, an early photograph of the quasar 3C48 yields an image almost indistinguishable from those of foreground stars. At bottom, a longer exposure by Allan Sandage reveals a faint surrounding glow—suggesting the quasar may, as several theorists suggest, be a young galaxy, viewed across a great distance in space and time. 3C48 lies four billion light-years away, so we see it as it was about one-fifth of the way back in time toward the creation of the universe; our own galaxy may have originated in a similar cosmic beacon.

ALLAN SANDAGE AND THE HALE OBSERVATORIES

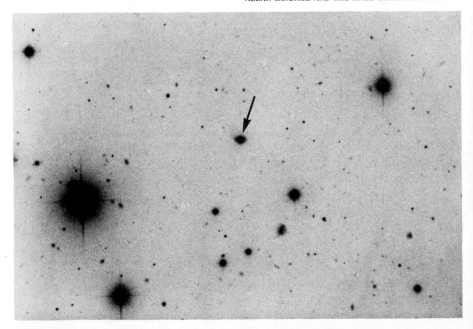

of scientific inquiry to regard observable effects as arising from 'causes unknown to science,' and this in principle is what creation-in-the-past implies." And Gold cautioned that continuous creation ought not to be rejected merely because it seems to defy common sense: "We often think we can make intuitive judgements on matters which have quite demonstrably never entered into our experience. There is hence no virtue in timidity in the choice of hypotheses. They may look drastic and offend our dwarfish preconceptions, so long as they do not offend any evidence. . . ."

Gold's remarks expressed something of the scientific climate in which the Steady State theory was created. In 1948, the expansion of the universe had been part of astronomy for almost twenty years but remained mysterious, insofar as the numbers it was couched in portrayed a cosmos too small and young to be believable. This situation invited radical theories; continuous creation of matter was hardly less credible than a Hubble cosmos younger than Earth. When Baade, then Sandage, corrected the mistakes in Hubble's numbers in the early 1950s, the climate changed. Advocates of the Steady State could no longer point to a paradox in the Big Bang theory, but had to present their case on its own merits, subjecting it to test by observation. These tests it failed.

Hoyle's predicted value for the deceleration parameter—the rate at which expansion of the universe is slowing—was not borne out by observation. A Steady State universe should contain many very old galaxies, some infinitely old; but as the 200-inch telescope was used to sample the cosmos to new depths, no extremely old galaxies were found. When Penzias and Wilson discovered the cosmic background radiation, fulfilling an almost forgotten prediction of the Big Bang theory, the Steady State received a mortal wound. By 1970 it had few advocates.

Enthusiasm for the Steady State, however, translated into enduring skepticism about Big Bang theory, which was attacked as a simplistic model that put blinders on its adherents. The trouble with the Big Bang theory, it was said, was that it encouraged astronomers to view data in its terms, to accept what fit and explain away what did not. "The guy who goes to the telescope at night and knows in advance exactly what he's going to observe, that guy drives me right up the wall," said Halton Arp, one of the dissidents. "Mavericks disturb things, rock the boat, cause trouble, and some, a majority, are dead wrong. But the ones who are right are very, very valuable. They know it is only from observation that you are going to learn anything new." So the Steady State-Big Bang debate evolved into a fresh version of an ancient question: To what degree does theory bias observation? On the one hand, some theoretical foundation must be admitted if a scientist is to make any observations at all; an astronomer who refuses to concede, say, the existence of stars probably will not get far. On the other hand, excessive adherence to theory can be crippling; Ptolemaic cosmology, which put Earth at the center of the universe, was accepted for centuries because it accounted for planetary motions reasonably well, while exceptions and inaccuracies could be explained away. The schism among cosmologists was as much a matter of philosophy as of science, and persisted when the Steady State theory faded away. In the 1960s it turned into a debate over quasars.

In 1959, thirty-three astronomers were polled about cosmology. "Did the universe start with a 'Big Bang' several thousand million years ago?" they were asked. A third said yes. "Is matter continuously created in space?" Fifty-four percent said no, 24 percent yes. "Is a Gallup Poll of this kind helpful to scientific progress?" All thirty-three answered no.

8

THE RED LIMIT

"O dark dark . . ."

—T. S. Eliot

ASTRONOMY PENETRATED TO NEARLY THE EDGE OF THE universe—or did so in the opinion of many astronomers—in the early 1970s. Newspaper accounts of the development raised as many questions as they answered—how can the universe have an edge?—but what had happened was perhaps less exotic than the phrase suggested. Big Bang theory predicted that there should be a distance in spacetime beyond which no galaxies or stars, no cosmic lanterns of any sort, appear. For years the prediction attracted little attention among observers because the distance to the hypothetical edge was far greater than that to which telescopes could photograph galaxies. Then in 1960 quasars were discovered. Redshifts in the light of quasars indicated that these objects were extremely far away and extremely bright, bright enough that existing telescopes could see them across billions of light-years. By the early 1970s quite a few high-redshift quasars had been found, but none with redshifts greater than a certain value, and that value corresponded with the time that the Big Bang theory implied galaxies first formed. It appeared that more distant quasars were not being seen because there were none. That cutoff point approached "the edge of the universe."

It is an edge not in space but in time. The Big Bang theory envisioned that the explosion, or whatever it was that started the universe expanding, was followed by an interval of time that elapsed before galaxies and stars began forming. Vast clouds of dust and gas organized into galaxies, which in turn subdivided into stars. The first light to grace the cosmos may have come from stars, from the furiously burning cores of young protogalaxies, or from some other source since extinct; the important point is that prior to the advent of this light, except for the heralding flash of the fireball, the universe abided in darkness. How much time darkness prevailed has proved difficult to estimate; theorists guessed it might have been as little as 10 million or as much as several billion years. In any case, astronomers today, probing cosmic history with their telescopes, ought to find that history exhaustible. If the first stars, quasars, or whatever, began shining, say, 15 billion years ago, then 15 billion light-years is the limit to which astronomers will be able to see anything. When they look farther they will encounter only darkness, because they will be looking back to a time before there *was* anything to see.

The edge should show up at about the same distance in all directions, for all observers anywhere in the cosmos. Nobody lives "at" the edge, or any nearer to it than anyone else, because the edge belongs to history. It is distant from us in the way the living Socrates is distant from us, as part of the irrevocable past. We can see deep into history because the universe is so large: Had Socrates, on the day of his trial in Athens, stepped aboard a spaceship and sped away at nearly the speed of light, we could look through telescopes today and see him, alive, on board that ship. His image would be a ghost, in the sense that it would be a long beam of light that might or might not still have a living man on the other end, but that is just another way of saying Socrates belonged to the past. His face, in the telescope, would be with us *now*.

So it is with galaxies. We see a young galaxy or quasar on the "edge" of the universe, as it was when its light first began to shine. In the billions of years since, presumably it has evolved into a middle-aged galaxy like our own, but we cannot yet see the result of that evolution, because it will take billions of years more for light to arrive with the news. The citizens of that evolved galaxy, turning their telescopes our way, will see the quasar or galaxy or protogalaxy or whatever our system was, billions of years ago, when the universe was young.

The four-dimensional universe of spacetime is difficult to imagine, because it is difficult, if not impossible, to visualize. If we try to fix a four-dimensional sphere or hyperbola in our mind's eye it eludes us, turns itself inside out. When we try to envision the universe of spacetime, at one moment it seems we are in the "center" of the cosmos, at the next that we are out near the "edge." To our perceptions either is correct, because there is no three-dimensional "center" to the universe, any more than there is a two-dimensional "center" on the surface of a sphere for the flatlanders. The question is not just "where" we are in the cosmos, but *when* we are, to which the Big Bang theory answers, we live roughly 18 billion years after expansion began, 15 billion years or so since the lights came on.

As the universe expands, observers in any galaxy find the other galaxies moving away at speeds proportional to their distance—the more remote each is, the faster it recedes—and this recession can be measured by determining the Doppler shift of light from the galaxy toward the red. (It is equally correct to imagine that the galaxy we observe is stationary and *we* are the ones moving so fast, since there are no fixed points of reference in the universe. Thinking about this aspect of it is one way to grasp four-dimensional cosmology.) Relativity theory requires that the speed limit is that of light: Remote galaxies may recede at velocities very close to that of light, but never equal to it. If the universe were infinite in time and space, the farther you looked the more galaxies you would see, and you could press on indefinitely and would always find more galaxies, with velocities ever closer to, but never achieving, that of light. In a Big Bang universe with a finite history, however, you would at some distance encounter the "edge" and see nothing beyond. In all versions of the theory, the edge is far away in space and time—lookback times on the order of 15 billion light-years are cited—and so will be found where galaxies are receding at something like 90 percent of the speed of light or more. Observing the putative edge of the universe therefore is a matter of seeing objects at great distances with large redshifts.

Existing telescopes could not record galaxies at that remove. For years the distance record belonged to Rudolph Minkowski, who in 1960, in a virtuoso performance on his last 200-inch observing run before retirement, obtained spectra for a cluster of galaxies receding at nearly half the speed of light, which would put their distance at some seven billion light-years in terms of lookback time. Electronic image-amplifying equipment later improved telescope performance, but the range to which galaxies could be observed remained less than half the estimated radius of the universe.*

That the universe should, early in its history, have thoughtfully provided cosmic beacons brighter than galaxies for the astronomers of today to observe seems an unexpected favor. But that is just what quasars appear to

* By "radius" I mean the time since expansion began, expressed in light-years, e.g., 18 billion light-years if the universe is 18 billion years old. The true radius of the universe will differ somewhat from this simple definition according to the degree of curvature of spacetime, which is in turn a function of the cosmic mass density.

be. Their redshifts indicate that virtually all quasars are remote. If so, they belong to an earlier cosmic epoch. Their tremendous luminosity suggests that they may represent a violent phase some galaxies go through when they form. Whatever the cause, the result appears to be that astronomers can make out quasars as far as the edge of the universe.

The discovery of quasars was made possible by radio astronomy. In the late 1950s, as new radio telescopes came into operation and the techniques for using them improved, radio astronomers compiled a succession of more accurate coordinates and refined maps of the sky. Many optical astronomers were interested in having a look at the radio sources, so much so that a lively traffic developed in "bootleg" radio coordinates, passed along informally before they were published. If an optical astronomer received a new set of coordinates from a friend at a radio observatory and aimed his telescope at the indicated spot in the sky, he stood a good chance of finding something unusual, to his delight but to the frustration of colleagues not favored with the information. In the competitive world of astronomy, where only a few dozen observers gain regular access to the large telescopes, friendships were cemented or eroded according to who was or was not favored with fresh radio plots.

In 1960, Tom Matthews, a radio astronomer at Cal Tech, was working on refining the coordinates of several radio sources observed by British astronomers at Cambridge and at Jodrell Bank, a radio telescope installation in Cheshire, England. These particular sources interested Matthews because each appeared concentrated, that is, limited to a very small patch of sky, which suggested that they might be far away. Matthews made up a list of ten such sources. Later all would prove to be quasars.*

Matthews managed to get an improved radio fix on one of the sources, 3C48 (for number 48 in the Third Cambridge Catalogue of cosmic radio objects). He passed the coordinates along to Allan Sandage, who used his next session on the 200-inch telescope to photograph the sky in the indicated spot. When Sandage developed the plate he found that the location of 3C48 coincided with a blue point of light of medium brightness, far too dim to be seen with the unaided eye but brighter than many of the things he was accustomed to photographing at Palomar. "I took a spectrum that night and it was the weirdest spectrum I'd ever seen," Sandage recalled. "I took the spectrograph off the telescope and put the photometer on, to check its colors, and the colors were different than any object I'd ever seen. The thing was exceedingly weird."

Several other compact radio sources were located with sufficient accuracy that optical telescopes could find them, and the list of known quasars grew to perhaps a half dozen. Each was analyzed with spectroscopes and the other tools of optical astronomy. None seemed to make sense. Several of the quasars showed clear spectral lines, but the lines fell in places along the spectrum that failed to coincide with the characteristic "fingerprint" of any known element. Equally difficult to explain was the fact that though

---

* The name, coined by a NASA physicist, stands for Quasi-Stellar Radio Source. It is doubly misleading—quasars are "quasi-stellar" only in the sense that they were at first mistaken for stars, and most are not radio sources—but by the time quasars were better understood it was too late to change it. Professional enthusiasm for the name never ran high, and when the Second Texas Symposium on Relativistic Astrophysics affirmed its use in December 1964, it did so by a vote of 20 yeas and 40 abstentions.

quasars looked like stars, they displayed bright emission lines in their spectra, instead of the dark absorption lines characteristic of stars. Emission lines had been recorded in galaxies that radiated powerfully in radio but were not usually found in stars. To add to the puzzle, Sandage, continuing his observations of 3C48, learned that its light varied in brightness over a matter of weeks.

The realization that the light of quasars was extremely redshifted eluded astronomers for two years. Most felt, with varying degrees of assurance, that because quasars looked like stars they probably were stars, even though previously no stars had been known to radiate powerfully in radio. "I tried very hard to identify the spectral lines of 3C48," Allan Sandage said, "but I was baffled. I had been raised to believe, with Hubble, that the only things that could have large redshifts were galaxies. So my mind was completely blocked. Both Tom [Matthews] and I assumed the quasars were stars."

A case study in this mental block was that of Jesse Greenstein, a transplanted New Yorker who had headed Cal Tech's graduate astronomy department since it opened in 1948.

Greenstein was something of a patriarch to the Mt. Wilson and Palomar astronomers; nearly half the observers winning time on the major California telescopes in the 1960s had trained at Cal Tech under his guidance. His favorite time of day was lunch in the Institute's wood-paneled, cubiform dining room, when he could sit and talk with his ex-students. "They're all my boys," he said in his rasping growl. "To me it's one of the finest, most exciting things in the world to sit at lunch with everybody and see my foster children being so bright. It's a wonderful place."

His specialty was stars. "I'm a person who understands

the spectra of atoms at various stages of temperature and pressure. A lot of things can be diagnosed about stars and galaxies from their atomic spectra and I'm something of an expert on it. I've studied, in particular and in great detail, how you get the composition of stars from their spectra. I can look at the spectrum of a star and tell you its age, its composition, its temperature and pressure. Any puzzle in spectroscopy I think of as a challenge." When Sandage, after trying for months to decipher the spectrum of 3C48 without success, showed it to Greenstein, Greenstein followed the bent of his training and interest and came up with a way to explain the quasar as a star.

"I kept talking to myself and finally convinced myself it was some kind of strange star, a thousand light-years or so away, the result of a supernova," he recalled. "I was able to find a physical condition in which atoms of helium and oxygen could be stripped of their outer electrons and produce spectra of ionized helium and five times ionized oxygen. You could understand these lines if you said these things resulted from a supernova."

Once, over lunch, Greenstein and a fellow Cal Tech astronomer, Maarten Schmidt, came achingly close to the solution when they talked about whether the lines could be those of familiar elements like hydrogen and helium redshifted to unfamiliar positions. But they dismissed the suggestion as, in Schmidt's words, "wild speculation." Later, when redshifts were found to be the answer, Greenstein was amused to reflect that he had treated quasars almost as a psychological projective test, stamping them with the imprint of his own preoccupation. "It was a classic story," he said, "of inhibition of one's creativity by knowing too much."

It was Schmidt who finally got at the truth about the quasar spectra.

One of the youngest of the Cal Tech astronomers, Schmidt had been born in Groningen, Holland, in 1929. He dated his interest in the stars from a summer night when he was twelve years old and looked through a small telescope at his uncle's home in Bussum. He soon made a telescope of his own out of a cardboard toilet paper roll and a magnifying glass he found in a junk pile. "It works, to my tremendous surprise," he wrote his uncle, and began observing each clear night with the telescope poked out of a third-story window. At Cal Tech, where he joined the staff in 1959, Schmidt was recognized as a skilled observer, someone who could extract top performance from a telescope. Virtually all astronomers do their own observing but few enjoy it; as Greenstein said: "When you're in the observer's cage of the 200-inch, the telescope turning and the stars going by, it's romantic, beautiful, marvelous, but on the other hand it's uncomfortable, tedious, bone-cramping, and you get sleepy." Schmidt, if he did not exactly love telescopes, at least felt at home with them. "Something about the sight of the 200-inch is deeply impressive," he said. "There's just something about it. Observing with any large telescope is essentially an adventure, even though your feelings usually are not exhilaration but of worry that you've forgotten something or are doing something wrong."

Schmidt began observing quasars in May 1962 and was able to obtain the first spectrum of 3C286. It revealed only a single line at a seemingly irrational location. He wrote for the *Astrophysical Journal Letters* what he privately called "a complaining note, saying, 'Here's this thing and we can't understand it.'" Another quasar, 3C147, turned up a few lines but they too defied Schmidt's attempts to explain them. 3C196 came out blank.

Then Sandage, acting on another of Matthews' radio coordinates, photographed the quasar 3C273 and discovered a thin jet, or spike, protruding from it. A similar jet had been found jutting out from the nucleus of M87, a galaxy in the Virgo cluster that Sandage had examined many times himself. M87 was one of the brightest known galaxies and was also a strong radio source. It was definitely a galaxy, and so 3C273, which resembled it somewhat, might be something like a galaxy. But first impressions die hard. Schmidt and Sandage did not imagine that the starlike quasars belonged in the distant habitats of the galaxies.

Schmidt obtained a spectrum of 3C273 and found a rich variety of lines that made no sense to him. He had just turned thirty-two.

On February 5, 1963, Schmidt sat in his Cal Tech office, a room with high ceilings, a gray wool carpet, tables buried under books and papers, and a large uncluttered oak desk. Sunlight filled the tall windows. He held the 3C273 spectrum, a sliver of film smaller than a stamp, pinched between the thumb and forefinger of his left hand. With his right hand he idly sketched the spectral lines on a yellow notepad. The British journal *Nature* had requested a note on the new quasar and Schmidt was preparing what he anticipated would be another inconclusive, "complaining" report. He felt vaguely irritable. He was coming down with a cold. He set down the pen, picked up a magnifier, and peered again at the frail spectrum on the film.

Greenstein walked down the hall past Schmidt's open door toward his own office. Before reaching it he heard

Schmidt call his name. Greenstein hurried back to Schmidt's office.

"Jesse," Schmidt said, "I think there's a 16 percent redshift in 3C273."

If it resulted from the expansion of the universe, that redshift would put the quasar out over a billion light-years.

Greenstein came round the desk. Schmidt showed him the pattern he had suddenly perceived, how four of the spectral lines were a well-known series of lines produced by hydrogen, moved 16 percent toward the red end of the spectrum. Intervening lines complicated the picture, but Schmidt could already see how most of them could be explained. The two men talked excitedly in the shorthand of astrophysics and spectroscopy.

Then a curious thing happened. Greenstein, who had studied the spectrum of 3C48 at length while developing his hypothesis that it was a star, felt that a vault door deep in his mind had swung open, revealing a chamber of the unconscious in which the redshift problem had already been solved without his consciously knowing it.

"Thirty-seven percent," he said.

"What?" They had been talking about 16 percent, the redshift on the spectrum in front of them.

"3C48!" In his mind's eye Greenstein could see the spectrum, suddenly rendered coherent. "Thirty-seven percent," he said to Schmidt, who was looking at him a bit oddly.

The two men went down the hall to Greenstein's office, took out the 3C48 spectrum from the files, and measured the displacement of the lines. It was 37 percent. The quasar, apparently, was nearly 4 billion light-years distant.

Looking back over the incident, Greenstein, a collector of Buddhist art with an interest in oriental philosophy, said,

"I have no belief in any mystic philosophy, but I have a great belief in the power of the unconscious."

There was a short celebration, then Schmidt returned to his desk and spent the rest of the afternoon reassuring himself that he had overlooked nothing, was not somehow making a mistake. By nightfall, he and Greenstein, satisfied that quasars did display large redshifts, had devised a preliminary proof that nothing else could account for the positions of all the spectral lines. The only way known for quasars to show such spectra was for them to be at great distance in an expanding universe.

Schmidt drove home and told his wife, in his Dutch-accented English, "Something horrible happened at the office today." He saw her shocked expression and realized he had chosen the wrong word. "I mean, something wonderful," he said.

"Your hope, when you start a career in science, is that you will discover something," Schmidt said ten years later. "Of course, there are many kinds of discovery. You may find a new comet in the sky, in which case your name goes on the comet and it's well established that you made a discovery, but that is actually an incidental thing, almost like being a witness to an accident. This quasar business was very different. It totally changed our insight and outlook upon the future study of the universe. Even now that we've been with quasars for years, we haven't recovered from the surprise. I simply can't believe it."

Whatever they were, quasars seemed to be some ten times brighter than galaxies. To search out quasars could be to probe the universe to unprecedented depth. As a staff astronomer at the Hale Observatories, Schmidt expected to be allotted twenty to twenty-five nights per year on the

200-inch. This time was his most valued resource. He invested most of it in hunting quasars. For the next three years—riding the telescope while it paced the stars and the music of Bach and Telemann played on the sound system in the dome—Schmidt accumulated specimens of quasars. By 1964 he was seeing farther in space and time than anyone had before. In 1965 he recorded quasars receding at up to 80 percent the speed of light.

Other astronomers joined in the search, among them E. Margaret Burbidge, a British astronomer who was one of the few women to take a place in the first rank of observers.* She and E. Joseph Wampler, inventor of an electrical image-amplification device she irreverently called the Wamplertron, measured the spectrum of a quasar receding at more than 90 percent the speed of light. At Kitt Peak Observatory in Arizona, C. Roger Lynds and Derek Wills identified a quasar redshift equivalent to 89 percent the speed of light, and in 1973, Mirjana Gearhart of Ohio

---

* Mrs. Burbidge, who was responsible for converting her husband, Geoffrey, from physics to astronomy, found when she arrived at Mt. Wilson that she could use the telescopes only by adhering to the fiction that she was his assistant. George Ellery Hale, annoyed by clashes between astronomers' wives at Yerkes Observatory before the turn of the century, had decreed that Mt. Wilson was for men only, and in the 1950s the tradition endured, as did others, including one by which the astronomer scheduled for the 100-inch telescope sat at the head of the table at mealtimes. Barred from Mt. Wilson's dormitory (known as the Monastery), the Burbidges stayed in the same cabin Hale had found atop the mountain when he first climbed it in 1903. Margaret Burbidge protested and was told the real reason for discouraging her was that the observatory had only a single bathroom. "Who cares about that in a dome?" she asked. "I mean, if it's only a matter of a *loo*. . . ." Ultimately she was admitted to equal status by the California observatories. When the Queen of England, in naming her director of the Royal Greenwich Observatory, severed the title from that of Astronomer Royal for the first time in its 300-year history, Burbidge resigned in little over a year and returned to the University of California at La Jolla. In 1971 she declined the American Astronomical Society's Annie J. Cannon prize for outstanding women astronomers, on the grounds that "because of the small number of women in the field," it would "not be surprising if we all in our turn are selected."

State and R. F. Carswell and P. A. Strittmatter at Steward Observatory in Arizona found a quasar, OH471, with a velocity of 90 percent the speed of light. Light from these objects may have been traveling, depending upon the model of the universe one chooses, for some 12 to 15 billion years.

The span of this odyssey did not especially impress those who achieved it. Accustomed to big numbers, the astronomers were more interested in the spectra themselves, in seeing unusual patterns of lines from above the high-frequency end of the spectrum shifted down into the visible region. Margaret Burbidge found it "terrifically exciting to think that we were actually seeing that far," not into the universe, but "into the ultraviolet—into regions of the spectrum normally blocked by the earth's atmosphere, that we thought we'd never see until space telescopes were developed." To Schmidt, this new spectroscopy was "for me much more exciting even than the discovery of the redshift of 3C273, though this has been appreciated by practically nobody. All those new lines had to be identified. It was a bootstrap operation. Nobody could tell me what lines to look for; the quasars had to tell it."

Talk of the edge of the universe arose not because quasars were so far away, but because none were found farther away. By 1973 redshifts had been measured for about 200 quasars, yet only a few of these were so distant as to be receding at 90 percent the speed of light and almost none were farther. If there were many quasars more remote, any of the major telescopes ought to have been able to photograph them. But as time passed and the list of observed quasars grew, the sharp cutoff at about 90 percent held, and a number of astronomers began to feel it probably was genuine, not an accident or an observational effect. "If the cutoff is indeed real," said the Hale Observatories annual report for 1971–72, "we may be looking back in time so far . . . we have seen 'the edge of the world' in space and time." *The New York Times* reported the story April 8, 1973, with the headline, "Men Report Seeing Edge of Universe."

Expansion of the universe began, according to measurements of galaxy distances and redshifts Sandage completed by 1975, 18 billion years ago, give or take 10 percent. Stellar evolutionists set the ages of the oldest observed stars in our own and nearby galaxies at 15 or 16 billion years. The level of the cosmic background radiation suggested a cosmos 15 to 20 billion years old. The quasar cutoff came at 16–17 billion light-years. Either the universe was born in fire some 18 billion years ago, or it was performing an adept impersonation of one that was.

But what are quasars?

Or rather, what *were* they? Few if any remain in the modern cosmos. In 1972 Schmidt and Sandage analyzed the accumulated data and found a pattern of quasar evolution with cosmic time. Locally, and therefore recently, no quasars were observed. Within a few billion light-years only a few were known. At distances such that their light had been traveling roughly half the history of the universe, 8–10 billion light-years, quasars were more common by a factor of 100. At still longer lookback times their frequency rose 300 times, then 1,000 times. Then at a point equal to about 90 percent of the age of the cosmos came the cutoff, and suddenly there were none. "Apparently quasars were numerous in the early universe," Schmidt said. "They evolved, died, and now are very rare. Probably they evolved into something else. I think they probably are the nuclei of galaxies, but mind you, that's speculation."

A task ahead was to understand the phylogeny of quasars, galaxies, and a bewildering array of other cosmic citizens. There were objects with names like Zwicky galaxies, Seyfert galaxies, N-galaxies, Haro objects, Braccesi objects, Markarian objects. Some were radio-prominent, some not. Some looked like peculiar galaxies, others resided in intergalactic space but might or might not be galaxies. Hubble and many since had attempted to piece together the evolution of galaxies with only the most preliminary success. Like the evolution of stars, it was a matter likely to consume the efforts of several generations. Just understanding quasars was going to be difficult.

One or two quasars were bright enough in the sky to be seen with an amateur astronomer's backyard telescope. If indeed they were halfway and more across the cosmos, the amount of energy they radiated into space must be enormous, as Schmidt realized the day he discovered the quasar redshift. Two astronomers specializing in infrared observations, Frank Low and Harold Johnson, measured the energy of a quasar in the invisible region just past the red end of the spectrum and found that it was many times more powerful than any known galaxy. Quasars were preposterously energetic; a few astronomers wondered whether they violated the known laws of physics. To shine that brightly for 10 or 15 billion years, each quasar would have to convert into energy more matter than exists in a large spiral galaxy. But in an evolving universe, the quasars do not shine brightly for anything like 10 billion years; as the Schmidt-Sandage analysis indicated, they were a feature of the early cosmos, after which they may have settled down to become the cores of what we call "normal" galaxies. They were no more required to keep up their prodigious energy levels throughout history than a sprinter is expected to run full speed all day.

A more serious problem in understanding quasars was that the brightness of many of them varied. Some flickered in a matter of hours. 3C446 was observed to double in brightness and then return to normal in a single day. 3C345 would sit quietly for three months at a time, then flare up. For an object to vary in brightness, there must be time for the mechanism involved in the change to affect the various parts of the object; atoms in one part of it must "know" what the other atoms want to do. The fastest this information can be transmitted is thought to be the speed of light. Therefore a quasar that pulses in, say, one day, can be no more than one light-day, or about 16 billion miles, in radius—if the pulse is intrinsic and not caused by some unknown observational effect. If the quasar's light source were larger it would be unable to keep its pulse in synchronization, somewhat like a marching band that cannot keep in time if the players scatter too widely. The discovery of variable quasars meant that quasars were not only shining brighter than whole galaxies, but were accomplishing this feat using a kernel of energy tiny by galactic standards.

To some astronomers the great distance attributed to quasars, their extreme brightness and their ability to vary in brightness over short periods made the whole theory of quasars look suspicious. Many of the skeptics were astronomers who earlier had been associated with the Steady State theory, and who kept alert when they felt their evolving-universe adversaries had climbed out on a limb. In quasars they perceived the issue they had been looking for. A debate arose that lasted years.

To attack the theory that quasars represented an early stage of cosmic evolution, it was necessary to challenge the assumption that the redshift in the light of quasars was caused by expansion of the universe, that quasars were far away, or "cosmological," in the parlance of astronomers. If the velocity was not imparted by expansion of the universe, where did it come from? James Terrell of the Los Alamos Scientific Laboratory suggested that quasars were small, relatively local objects that had been ejected from the core of our galaxy, and were moving away rapidly only because the Milky Way had spewed them out. Terrell's proposal sparked some interest on the part of Fred Hoyle and Geoffrey Burbidge, who wrote a paper elaborating on it. The trouble was that if our galaxy was throwing off quasars, presumably some other galaxies ought to be hurtling toward us, in which case their spectra should be shifted toward the blue instead of the red. No blueshifted quasars have been found.

Another possibility was to say the redshifts were not caused by velocity at all but were gravitational. In relativity, light is redshifted as it climbs out of the spacetime "well" of the star it is leaving. But quasar redshifts are large, and a gravitational field of the required intensity implied pronounced side effects. In a paper that even the opposition described as "overwhelming," Schmidt and Greenstein showed that the expected side effects were not present. They went on to exclude almost any non-velocity explanation for the redshifts.

Hoyle found himself somewhat at cross currents on the issue. Just before Schmidt discovered quasar redshifts, Hoyle and William Fowler of Cal Tech had published a paper suggesting that the cores of powerful radio galaxies were massive collapsing "superstars." This theory stood to gain some support if quasars represented an early stage in the development of galaxies. On the other hand, any suggestion that the universe was once different than it is today—was once full of blazing quasars—ran counter to Hoyle's Steady State theory, and though Hoyle was, by his own account, "much less interested" in the Steady State cosmology than he had been, he retained a certain sympathy for it. Consequently, he was disinclined to accept the view that quasars lay at cosmological distances. He called it "a fifty-fifty situation" whether quasars were really far away.

The burden of the skeptical position on quasars was instead taken up by a younger astronomer, Halton Arp.

The son of a commercial artist, Arp grew up in Greenwich Village and Woodstock, New York. "I had a pretty chaotic childhood, so I sort of gravitated toward a more ordered, academic existence," he said, but when he joined the Hale Observatories staff, some of the other astronomers there came to feel his approach was as much art as science. Arp was fascinated by the exotic and unusual, of which there is a great deal in the sky. He moved into a Moorish stone house a magician had built on a Pasadena hillside, and compiled his *Atlas of Peculiar Galaxies*. It bulged with photographs, some from observatory files and others he took himself at Palomar, showing odd galaxies that caught his fancy—distorted spirals trailing like birds with broken wings, giant systems with veils of glowing gas spread between them, smoke rings, strings of beads, warped lumps like bubbling yeast. Each was home to billions of stars. Arp chose them with an artist's eye, picking what looked interesting.

Arp's approach contrasted with that of Sandage and others who worked in the tradition of Hubble. Sandage devoted years to determining just how bright, on the average, the dominant galaxy in a typical cluster of galaxies was, so that he could then use that knowledge to estimate the distances of more remote clusters; he sought constancy among the galaxies. Arp was after novelty.

One winter night in 1966, thunderstorms on Palomar prevented Arp, scheduled to use the 200-inch, from observing. He sat up all night in the observatory dormitory, drinking coffee, listening to rain wash over the roof, and hoping the skies would clear. To pass the time he paged through a catalogue of cosmic radio sources. Radio sources were a sore point with him. He resented radio men who had passed unpublished coordinates along to Sandage and Schmidt instead of him, enabling them to discover and explore the quasars before Arp knew what was happening. "If you got hold of one of those radio positions," he said, "all you had to do was sit in one of the big telescopes, take a photograph, take a spectrum and write a paper. It was great. But I couldn't get hold of any of those top secret positions. I was elbowed out of the field, even though my work was right at the nexus of the study of the galaxies." As Arp looked through the list of radio sources, he thought about a letter he had received from a friend, J. L. Sérsic in Argentina. Sérsic had noticed that several compact radio sources were located in the sky near the position of galaxies from Arp's *Atlas;* he wondered if the proximity were purely coincidental. Arp wondered so also. He began comparing the celestial coordinates of radio sources in the catalogue with those of his favorite strange galaxies.

By 4 A.M. Arp had come up with several radio objects located in the sky near peculiar galaxies. "The only thing I didn't know was just what the radio sources were," he said. "In the morning I went over to the dome and looked them up in the literature. Three or four of them turned out to be quasars." In the following weeks Arp assembled a list of twenty-nine galaxies, most from his *Atlas,* whose position in the sky was adjacent to that of quasars or other radio sources. Arp felt that the galaxies and the quasars did not just happen to lie along the same line of sight but were in fact physically associated. If so, the quasars could not be as far away as their redshifts indicated but had to be at about the same distance as the galaxies. The galaxies were all relatively nearby, as intergalactic distances go, and so, if Arp was right, the quasars were nearby too.

More to the point, Arp obtained what he felt was photographic evidence of quasar-galaxy interactions. In several cases, galaxies with disturbed or unusual-looking nuclei displayed filaments of hydrogen gas pointing at a nearby quasar. In others, two compact radio sources appeared along a straight line to either side of the galaxy nucleus. Arp felt these photographs constituted evidence that quasars were ejected from the centers of galaxies.

Astronomers opposed to the Big Bang theory were happy to hear of Arp's work. The Big Bang had gained popularity until it threatened to dominate cosmology. Arp's papers attacked a central assumption of the theory, provoked questions, opened new lines of debate. Many found this healthy for the field in general. The Steady State theory might be dead, but that did not mean the Big Bang theory was correct. The universe is full of surprises; maybe the true nature of quasars was going to be one of them. In papers supporting Arp, the odds that the apparent proximity of quasars and relatively nearby galaxies he found could be coincidence were variously cited at one in ten, one in five hundred, and one in a thousand.

Widely reported in the popular and semiscientific press, Arp's position generated considerable public interest. When the controversy had not cooled after five years, a debate was arranged, reminiscent of the Shapley-Curtis debate of 1920. Arp's opponent was John Bahcall, a young physicist at the Institute for Advanced Study in Princeton. Arp had taken to talking of "discordant redshifts." When he found a galaxy and a quasar that appeared to be associated, but the quasar had a much larger redshift than the galaxy, he concluded that something was wrong, or "discordant." Bahcall and his wife, Princeton University physicist Neta Bahcall, had identified a catalogued quasar with a surrounding cluster of galaxies that shared a similar redshift, a relationship they termed "concordant" in faint parody of Arp. Arp had a look at the object, and in a bristling response, charged that it was not a quasar at all. Within a few months James Gunn, a friend of the Bahcalls and, unlike them, a trained astronomer, photographed an indubitable quasar associated with a cluster of galaxies of equal redshift. This observation was an important one for the quasar orthodoxy because it supported the view that quasars really were at the distances indicated by their redshifts. Finding such missing links—clusters containing both galaxies and quasars—was difficult because galaxies were too dim to see clearly at distances of more than a billion light-years, while quasars didn't begin to proliferate until two billion light-years or more. Arp argued that a few such concordant examples did not matter; if he could find just one case of a low-redshift galaxy associated with a high-redshift quasar, and prove to everyone's satisfaction that it was genuine, then the foundations of quasar research would be uprooted.

Arp and Bahcall met under the auspices of the American Association for the Advancement of Science in Washington, D.C., December 30, 1972.

Arp brought along his beautiful photographs of galaxies, prints showing alleged bridges of gas and dust between objects which, by the conventional interpretation of red-shifts, ought to be millions or billions of light-years apart. He had broadened his original thesis into a general attempt to find any combination of cosmic objects that appeared to be associated yet had differing redshifts. His position amounted to an assault on the dependability of the redshift-distance relationship, a key to deep-space astronomy for the forty-five years since Hubble discovered it. He paraded slides before the audience: Here is a galaxy with a distended arm that points toward a quasar, here is one linked to another with twice its redshift, here is a spiral clearly perturbed by the gravitational field of another that Bahcall says must be far away from it. Arp offered no real explanation of what caused quasar redshifts, if not expansion of the universe. His point was simply that something unexplained was going on.

Bahcall, a young man with a preoccupied manner, took the dais. He recounted the accomplishments of Hubble, Humason, Sandage and other cathedral-builders of modern astronomy, then bore down on what he saw to be the central weakness of Arp's reasoning: Arp, he said, kept talking of "associations" between cosmic objects with discordant redshifts, but never defined what an association was. His rules changed to fit each case. "The skies when photographed with large telescopes reveal so many individual objects on any photographic plate that one can find almost any configuration one wants if one just hunts," Bahcall said, "even stars arranged as four-leaf clovers." Bahcall charged that some of what Arp perceived as intergalactic bridges were observational errors, while other "associations" resulted from a distant object falling by chance along the same line of sight as a nearby one. As for the statistical arguments by Arp and others that quasars too often appear close to peculiar galaxies to be accidental, Bahcall cautioned against statistical cases made after the fact. A photograph of Times Square may show six teenage girls in a line, or a man from Borneo arm-in-arm with a woman from Paris, and the odds against finding such a thing may be high, but the odds of having found it that one time are always 100 percent, because there it is.

In rebuttal, Arp defended both his observations and statistics. "Obviously there is always some finite chance that any one association could be accidental," he said. "If each association is considered separately, each could be dismissed on these grounds. But what is the chance that two or three or a half dozen could be accidental?"

When the debate was over, it was difficult not to be impressed by Arp's sincerity and his love for the mysterious galaxies he studied, but it was also difficult to feel that his case had suffered anything short of demolition. Lacking a solid theory of what quasars were, he was really saying that more orthodox astronomers, the men who had "elbowed" him aside, did not know everything. That was certainly true but it did not change the fact that quasars, by the weight of the evidence, really did seem to be astonishingly far away.

To the extent that Arp did have a theory, it was that the nuclei of galaxies may explode and throw off quasars. This capitalized on the idea of exploding galaxies, something of a fad in astronomy in the 1960s. A number of galaxies with bright centers had been found, many of them strong radio objects, and in 1961 Sandage photographed one, M82, with a special filter and was astonished to see what looked like a galaxy blowing itself apart. Galactic suicide was discussed for years, with a few theorists going so far as to attempt to estimate our chances for survival should the

Milky Way detonate, but in 1976 Philip Morrison and a colleague established that M82 probably was not exploding at all. Violent galaxies may well exist, but whether any are so violent as to eject large quantities of matter into space remained to be seen.*

The quasar dispute left rifts in the community of astronomers. Arp and his sympathizers were called spoilers and headline-grabbers by less impulsive colleagues, who in turn were perceived as imperious and egotistical. Sandage and Arp, who had been close friends, were no longer speaking. They worked in offices fifty feet apart in Pasadena, assaying cities of stars in a language of mathematics and physics not more than a handful of individuals fully understand, and did not talk to each other.

"I don't talk cosmology at the office any more," Arp said one dusk, at home, as he watched the sun set. "It's a very touchy subject. The people committed to the other side take it pretty seriously. Part of it is ego. I have a friend who says, 'Any man who's egotistical enough to think he can figure out the whole universe is pretty damn egotistical.' The other part is that the success or failure of your theories gets associated with the success or failure of your scientific

---

* The M82 episode marked one of two occasions when the normally prudent Sandage erred. The other came in 1965 when he wrote a paper titled "The Existence of a Major New Constituent of the Universe: The Quasi-Stellar Galaxies." Using a two-color survey technique he had developed with Ryle, Sandage found many quasars that did not radiate radio noise. The essence of the paper was sound, in that radio-quiet quasars do outnumber radio-prominent ones, but Sandage overlooked the fact that many of the objects he took for "quasi-stellar galaxies" had already been studied by Cal Tech astronomer Fritz Zwicky, among others. Zwicky found that they were compact galaxies that probably had a place in quasar-to-galaxy evolution, but hardly deserved the label "major new constituent of the universe," since Zwicky already had observed them. Sandage was apologetic. "I rushed in too quick," he said.

career, and most people here have their whole lives wrapped up in their careers.

"You know, a lot of us, growing up, wanted to be famous scientists. At Cal Tech we struggled to learn very complicated things. Our lives, our success, depended upon whether we could pass those exams, regurgitate stuff like Hubble's redshift-distance law. Now, in later years, having learned it has become almost a part of our security. People take it seriously and if somebody questions its validity, their psychological security gets shook up.

"It's so competitive these days. Pressure on the telescope, people scrambling for time—that's the gut issue, who gets what amount of time on the 200-inch. Every time you go down to Palomar you feel you have to make an astounding discovery or you're in peril. I think graduate school failed to get across to everyone the idea that it would be all right to have a pluralistic society, astronomers working on different things and coming up with different results. I think that failure is unfortunate.

"Sooner or later we will find out what quasars really are, but whether we find out how to communicate with and trust one another, well, that seems to me much more important."

Sandage withdrew into deepening isolation. He turned up less often at scientific conferences. At one meeting Geoffrey Burbidge, in a jovial mood, drew a horizontal line on a blackboard and with the help of the audience charted astronomers along the line according to whether, and how strongly, they felt quasars were denizens of early cosmic history. Arp was awarded the extreme left wing. Schmidt, who was devoting almost all his time to quasars on the assumption that they were far away, was placed at the far right. Margaret Burbidge asked to be put slightly left of center. Bahcall went to the right, just shy of Schmidt. Where, someone asked, do we put Sandage? Burbidge thought for a moment and then, to general laughter, wrote in Sandage's name on the extreme right but high above the others. Sandage was above it all. He had not come to the conference. He was back in California, trying, as he sometimes put it, "to rewrite the book of Genesis."

In his office, where a portrait of Eddington hung on the wall, and every level surface was stacked feet deep with photographs, technical papers and computer printouts, Sandage said, "It doesn't matter whether there is a controversy over quasars. It doesn't matter what anyone says. The ultimate judge is nature. The majesty and excitement of the thing is just that, that it is independent of man, *and it is out there!* The galaxies will go on regardless of the councils of the scientists. There is an *ultimate truth!*"

He pushed a redshift-distance diagram across the desk. *"This,"* he whispered. "This is what nature *says!"*

9

THE FATE OF
THE UNIVERSE

*The explorations of space end on a
note of uncertainty. . . . We measure
shadows . . . we search among
ghostly errors of measurement. . . .*

—EDWIN HUBBLE, 1935

SANDAGE DESCRIBED HIS LIFE'S WORK AS "A SEARCH FOR TWO numbers." One was the rate at which the universe is expanding, known as the Hubble constant, designated H_o. The other, the deceleration parameter, represented to what degree the expansion was slowing down. The deceleration parameter was symbolized q_o. If these two numbers were known they could establish a great many things about the universe, including whether it will expand forever.

Revisions in the Hubble constant over the years were so radical as to be something of an embarrassment; one astronomer wrote that they played "in the history of our subject much the same part as the crash of 1929 in the history of Wall Street." With each revision the universe was perceived to be larger and older, and therefore to be expanding at a more gradual rate, than had been thought before. When Sandage was a graduate student at Cal Tech, the Hubble constant was being quoted at about 100 miles per second per million light-years. That is, for every million light-years farther out one looked, one would observe galaxies receding at an additional hundred miles per second velocity. Baade's work on Cepheid variable stars cut this figure in half; in 1956, astronomers on three continents reduced it further; Sandage scaled it down still more in the 1950s and again in the 1970s, until the "constant" wound up at 10 percent its original value. According to Sandage's 1976 figures, galaxies recede at ten miles per second per million light-years distance. Sandage claimed a margin of error of under 15 percent. "It's a real number now," he said. This would make the age of the universe (meaning time since expansion began) 18 billion years, give or take three billion.

Determining the rate the universe expands consumed first part of Sandage's time, then most of it, over a period of twenty years. He had to obtain thousands of redshifts and find accurate distances for each galaxy by some means independent of redshift as well. The first chore was not especially difficult but the second was. To determine galaxy distances Sandage elaborated upon the stairstep procedure of Shapley and Hubble. First he rechecked distances to relatively nearby galaxies by isolating Cepheid variables and other recognizable types of stars in them; the absolute magnitude of these stars was known from studies of them in our own galaxy, and by the classic astronomical method of comparing their absolute magnitude with their apparent brightness in the sky, distances could be judged. This first step took Sandage out to about 5 million light-years, the limit to which the 200-inch telescope could make out Cepheid variables. To go farther, he examined certain very bright components of the nearby galaxies, primarily giant blue stars and knots of glowing hydrogen gas; when he knew the intrinsic brightness (and in the case of the gas knots, size) of these features with confidence, he then could estimate the distance to any galaxy in which they could be photographed. This step was useful to 30 million light-years.

Beyond that, Sandage shifted from using stars in galaxies to using entire galaxies as his "standard candles." This procedure was made feasible by the remarkable fact, discovered by Hubble, that clusters of galaxies resemble one another. Each cluster typically contains one giant elliptical galaxy or one dominant spiral, and so forth, and a type specimen in one cluster is likely to be of about the same brightness as its counterpart in each other cluster. By studying many clusters of galaxies Sandage was able to learn to what extent this grand-scale display of the uniformity of nature could be trusted. He then used these standard galaxies in clusters of galaxies to measure distance as once he had used stars. He penetrated well beyond 300 million light-years, sampling enough of the universe to feel assured that he was measuring expansion unfettered by local deviations. (The expansion does not occur within clusters of galaxies, which are bound together gravitationally, but out between clusters, and since there is reason to suspect that the clusters themselves are organized into metagalaxies, it is necessary to observe a large number of galaxies to be sure the observed expansion rate is really typical of the cosmos and not slowed by the proximity of clusters.) A number of other astronomers contributed to this work, although Sandage dominated the field. A variety of other distance-measuring techniques were employed, including apparent angular diameters of galaxies and their radio profiles. The idea was to try as many independent methods as could be found in order to reduce the margin of error, to mount "enough lines of attack to over-determine the parameters," as Peebles at Princeton put it.

The result, when redshifts were plotted against distance, was the straight line of a uniformly expanding universe.

Gathering the data was dry toil that Sandage commenced in youth and concluded in middle age. It held little glamor for him. Most mornings he had to force himself to sit at his desk and resume work. "Self-discipline," he liked to say, "is the highest human virtue."

The second number, the deceleration parameter, remained.

Standard Big Bang theory assumes that the expansion of the universe must slow down in time, since all the energy of expansion was imparted at once, in the beginning, and mutual gravitational attraction among the galaxies over billions of years should act as a brake on it. The question is, at what rate is it slowing down? If the deceleration is sufficient to one day halt expansion altogether, collapse of the universe should follow; if so, the universe is "closed" in the sense that both its past and its future are finite in time. If deceleration is less than the critical value, expansion will continue forever and the universe is "open," its future infinite in time. The closed universe may be called spherical or elliptical in the language of the non-euclidean geometries. The open cosmos is hyperbolic. The models of course are four dimensional—one dimension of time plus the three dimensions of space—which is why the spacial geometry of the universe presumably can be determined if we know its fate.

A deceleration rate sufficient to close the universe will show up as a departure upward from the straight line on a chart of the Hubble velocity-distance relation. The departure appears because at high lookback times we are seeing a period in cosmic history when the universe expanded more rapidly than it does today; the difference between the velocity of expansion then and now yields the rate of deceleration. The universe, however, is vast, and the Hubble relationship must be tested to extreme distances before any deceleration rate is likely to become evident. Spectra of galaxies billions of light-years away are difficult to record and analyze, and independent checks on distance are fraught with error. From the outset Sandage remained pessimistic about ever being able to establish the deceleration parameter by studying galaxies with earthbound telescopes.*

In 1972 Sandage attempted to extract it from the data he and three other astronomers had accumulated in the course of refining their value for the Hubble constant. Since the effect of the deceleration parameter was expected to be subtle—smaller, in fact, than Sandage's margin of error—he used a computer program designed to eliminate "noise" from the data and seek out an underlying trend. The results suggested a closed universe, but were highly uncertain. "No decision is yet possible . . ." Sandage wrote, "as to whether the universe is open or closed."

An alternative was to look for the deceleration rate not in gross numbers at great distance, but in subtle trends closer to home. Clear-cut expansion of the universe, or "pure Hubble flow" as it was sometimes called, begins about 300 million light-years away, where the Local Group's gravity is no longer an important inhibition. Working with G. A. Tammann of Switzerland and Eduardo Hardy of Chile, Sandage studied the recession rate of galaxies lying just beyond that distance. Thirty clusters of galaxies were sampled, their redshifts and distances

---

* The era of constructing large optical telescopes on Earth is drawing to a close, and observers' hopes for the future involve orbiting large instruments in space, where there is no atmospheric interference and a broader portion of the spectrum can be observed. A proposed 119-inch automated reflector in orbit could record objects a hundred times fainter than the limit of the 200-inch on Mt. Palomar. In the meantime the trend has been toward building more telescopes in the 100- to 200-inch range, particularly in the southern hemisphere, and equipping them, as well as older telescopes, with electronic image-enhancing gear and computers. The Soviet Union's 263-inch reflector in the Caucasus Mountains is likely to remain the largest telescope on Earth for some time to come.

measured with the greatest possible accuracy. All fell almost perfectly on the flat Hubble line. There was no perceptible tendency toward deceleration.

That the universe expands homogeneously in all directions was well established, but Sandage had not expected expansion, even locally, to be *that* homogeneous. The thirty clusters showed *no* concerted variation from uniform expansion. Several giant aggregations of galaxies, including the rich Virgo, Coma and Ursa Major clusters, lie not far, intergalactically speaking, from the clusters Sandage studied, and he expected their gravity to have slowed expansion locally. Instead, all the clusters appeared to be flying apart from one another, and from ours, at an unfettered rate. This indicated that the universe was expanding with almost unchecked vitality, that the deceleration rate must be small.

More evidence to the same effect came when improved studies of the stars in globular clusters, previously thought to be on the order of 11 billion years old, indicated they were more nearly 14 billion years old. The older the cosmos, the more powerful the Big Bang must have been to result in the rate of expansion we see today, and so the less likely it is that deceleration is enough to close the universe. The new globular cluster age estimates also fit fairly well with Sandage's value of 18 billion years since the Big Bang, a number obtained by assuming the universe has been expanding without significant deceleration since genesis.

In 1974 four younger astronomers—J. Richard Gott III and James Gunn of Cal Tech, and David N. Schramm and Beatrice Tinsley of the University of Texas—wrote a long paper summarizing evidence that the universe is open and will expand forever. In contrast to the lofty, somewhat lonely tone of Hubble and Sandage's publications, their paper was the work of a team that drew on the work of sixty-four astronomers, and it had a wry, ironic tone compared to that of the cosmological cathedral-builders who had preceded them. They began with a quotation from Lucretius: ". . . In all dimensions alike, on this side or that, upward or downward through the universe, there is no end." They approached the question of the fate of the universe in terms of matter: Does the cosmos have enough matter in it to generate the gravity needed to stop expansion? Evidence from a variety of sources suggested it does not.

If the estimated mass of all galaxies is added together, the Gott group noted, the total is too little by a factor of ten to close the universe. One could propose that there is a lot of dust and gas out between galaxies, but our ability to see clearly for enormous distances—some 10 billion light-years if the quasar redshifts are cosmological—suggests that intergalactic space is clear. The Gott group considered several suggestions about where large amounts of matter might be hidden, and rejected nearly all of them. Where could the "missing matter" be? "Loopholes in [our] reasoning may exist," they wrote, "but if so, they are primordial and invisible, or perhaps just black." Black, that is, as a black hole.

Insular, excommunicant, perpetually hungry, a black hole is the perfect place to hide anything. It is a well in spacetime so steep that nothing, not even light, can climb out of it, a confluence of the great in small, in circumference smaller than the city of Boston, in diameter infinite—you can feed a tape measure in forever and the black hole will simply eat it. It will eat hydrogen gas, planets, flowers or kings with equanimity. Each, falling in, departs our universe for good. The black hole will tell us nothing about its victims. It will tell us, in fact, nothing about itself except its mass, electrical charge and angular momentum. "A black hole has no hair," John Wheeler liked to say.

That these baleful sinks exist was a prediction of relativity. Karl Schwarzschild, one of the few who realized, prior to Einstein, that the universe might be viewed in terms of non-euclidean geometry, went to work soon after general relativity theory was published in 1916, investigating the geometry of spacetime in the vicinity of massive objects, which practically speaking meant stars. His calculations revealed that for a star of any given mass there is a "Schwarzschild radius" such that if the star is compressed to a size smaller than that radius, it will collapse to so high a density as to pinch itself off from surrounding spacetime. The result is a kind of hole in space. Things can fall in but nothing gets out.

The Schwarzchild radius became part of theoretical astrophysics, but whether a real star ever suffered such a fate was another matter. Densely packed white dwarf stars were known to exist and were believed to result when stars burned out their nuclear fuel and collapsed, the outward pressure of their light and heat no longer able to balance the grip of their own gravity. White dwarfs were fantastic objects—a cubic inch of their surface could weigh tons—and the possibility that stars might collapse even further was not taken very seriously at first. Then in 1930 a young Indian student, Subrahamanyan Chandrasekhar, calculated that if a dying star were much more massive than the sun, it would collapse right through the white dwarf stage to become something even more compressed.

Chandrasekhar was studying under R. H. Fowler at

Cambridge at the time, and Fowler brought his work to Eddington's attention. Eddington saw that if the calculations were correct, a sufficiently massive collapsing star could become so compressed that it sucked in its own light and cut itself off from the rest of the universe. This struck Eddington as ridiculous, and he felt he had discredited Chandrasekhar's theory by reducing it to absurdity.

In America the nuclear physicist J. Robert Oppenheimer, urged on by, among others, George Gamow, looked into the physics of gravitational collapse, and with several students published two important papers in 1939. The first theorized that stars could collapse through the white dwarf stage and become "neutron stars," objects whose intense gravity broke down the structure of atoms within them. The term was borrowed from Baade and Zwicky, who five years earlier had suggested that exploding stars might leave neutron stars behind as a sort of ash.* The second Oppenheimer paper discussed the possibility of a star crushing itself beyond the observable limit. "When all thermonuclear sources of energy are exhausted a sufficiently heavy star will collapse," the paper began. "Unless fission due to rotation, the radiation of mass, or the blowing off of mass by radiation, reduce the star's mass to the order of that of the sun, this contraction will continue indefinitely. . . . Light from the surface of the star is progressively reddened, and can escape over a progressively narrower range of angles." As the gravitational redshift approaches infinity, the star disappears and we have a black hole.

Interest in black holes reawakened in the 1960s with the

---

* They were vindicated thirty years later with the discovery of a pulsar, or radio-bright neutron star, in the Crab Nebula, where Chinese astronomers had recorded a supernova in the eleventh century.

discovery of quasars, the great energy of which might be explained by gravitational collapse, and of pulsars, which were shown almost certainly to be neutron stars. If there are things in space strange as neutron stars, one reasoned, why not black holes? Theoretical models showed that while finding solitary black holes in space by optical means might be impossible, they could be detected if they were part of double-star systems: the black hole would suck surface gas off its companion star, and the gas, whirling down into oblivion, would emit a flood of X-rays.

Their exotic name aside, X-rays are just another part of the electromagnetic spectrum, occupying the zone between ultraviolet light and gamma rays. They cannot be observed reliably from Earth because the atmosphere tends to absorb them. An Italian-American X-ray satellite, Uhuru, launched in 1970, eventually located over 100 cosmic X-ray sources, of which some were other galaxies, some neutron stars, and one, Cygnus X-1, was possibly a black hole. The issue remained in doubt, but several researchers pronounced themselves almost certain they had found a genuine black hole. Cygnus X-1 appeared to be feeding off its companion, the supergiant star HDE 226868, in a gravitational web so tight that the two whipped around their common center of gravity once every 5.6 days. If that model were correct the system would be quite a sight—the tormented supergiant pouring glowing gas across space and into an invisible spacetime vortex—but interstellar cosmonauts might be forgiven for passing up a close look. The crew of a spaceship falling into a black hole would experience, in their last moments, a growing redshift of starlight resembling the death of the universe, though it would be of course their own.

Although scientists were to argue for years whether it and other candidates located thereafter really were black holes, public response to discovery of Cygnus X-1 was enthusiastic. Science fiction writers made room for black holes in stories, poets heralded them, a few painters accepted what may be the ultimate technical challenge and attempted to paint their portrait. California physicists considered the feasibility of black hole power plants. Soviet theorists proposed that a massive, rotating black hole might constitute a bridge between two distant parts of the cosmos, one a daring pilot might navigate. In Texas two physicists proposed, deadpan, that a tiny black hole caused the Siberian forest explosion of June 30, 1908, when a brilliant fireball of uncertain origin slammed to earth near Tunguska, 100 miles north of the Sea of Okhotsk, and leveled 300 square miles of trees. They imagined that the black hole continued through the earth and emerged from the North Atlantic between Newfoundland and the Azores. They urged ships' logs of the period be searched for sightings of the reappearance.

As Kip Thorne of Cal Tech wrote, "Of all the conceptions of the human mind from unicorns to gargoyles to the hydrogen bomb perhaps the most fantastic is the black hole."

To cosmologists, the significance of black holes lay in their potential role as lockers for hidden matter. As Gott and others showed, all the known matter in the cosmos is not nearly enough to stop expansion. Unless ten times more matter is found, the universe is open. Could 90 percent of the stuff of the universe be locked away in black holes?

If all black holes were formed in the collapse of stars, probably not. If massive stars have been dying at a constant rate since the galaxies formed, a spiral system the size of the Milky Way ought to contain today about 100 million black holes. To avoid erring on the low side, let us say that stellar collapse was once more common than it is today and boost the estimate by a factor of ten, to one billion. Assume, again on the high side, the average black hole mass is ten times the sun's. (A more likely estimate is half that.) The result is to add 10 billion solar masses to our galaxy. That is a lot of mass, but it makes the galaxy heavier by less than 5 percent. We can go on to assume that every galaxy once had a quasar at its center and that the quasar remnant remains as a black hole. Theoretically this giant black hole could have a mass of up to one billion times the sun's. Even if we double that, we are still a long way from finding enough mass to stop expansion.

Gott and Gunn considered these and other "missing mass" theories, discarded them, then concluded, by studying the gravitational interaction of galaxies in clusters, that galaxies simply do not act as if they are ten times more massive than currently estimated. *"Galaxies themselves cannot close the universe,"* the Gott group wrote (their italics). If black holes harbor most of the mass of the cosmos, they must not be in galaxies.

The Gott group concluded that the universe probably, though by no means certainly, is open. "A clear verdict is unfortunately not yet in, but the mood of the jury is perhaps becoming perceptible," they wrote. Sandage, who previously leaned toward a closed universe, shifted toward their view, prompted by the globular cluster star ages and his own studies of expansion rates in nearby galaxies. "The conclusion that the universe will expand forever seems inevitable," he wrote in 1976.

If so, it is destined for heat death. Eventually all the stars will go out and the material for making new ones will have been exhausted. The universe will be an entropic duchy, carting black galaxies outward in a black cosmos forever.

If not, expansion will one day stop, contraction follow, and a new cosmic fireball be born. Present evidence weighs against this but is not conclusive. As John Wheeler said, "It's too soon yet to say what the story is. The excitement is just beginning. Nobody should throw in his hand at this point."

Whether the cosmos ends in fire or ice, apparently it will end, which is to say that it will change into something we would not recognize and could not live in. Some find this depressing. Bertrand Russell regarded it as tragic that "all the noonday brightness of human genius ... the whole temple of Man's achievement must inevitably be buried beneath the debris of a universe in ruins. . . ."

But later he cheered up a bit, as might we all for various reasons. We have little more personal stake in cosmic destiny than do sunflowers or bottleflies. The transfiguration of the universe lies some 50 to 100 billion years in the future; snap your fingers twice and you will have consumed a greater fraction of your life than all human history is to such a span. Life is not edifice but process. We owe our lives to universal processes, in that identity of cosmic and earthly destiny glimpsed by George Ellery Hale, and as invited guests we might do better to learn about them than to complain about them. If the prospect of a dying universe causes us anguish, it does so only because we can forecast it, and we have as yet not the slightest idea why such forecasts are possible for us. A few figures scrawled on a piece of paper can describe the rate the universe expands, reveal what goes on inside a star, or predict where the planet Neptune will be on New Year's Day in the year A.D. 25,000. Why? Why should nature, whether hostile or benign, be in any way intelligible to us? All the mysteries of science are but palace guards to that mystery.

10

EXPANDING UNIVERSE OF THE MIND

What led me to my science and from my youth filled me with enthusiasm, is the fact—not at all self-evident— that our laws of thinking conform with the lawfulness in the passage of impressions which we receive from the other world, thus making it possible for man to gain information about that lawfulness by mere thinking.

—MAX PLANCK

As I have been telling the story, in this century humans first saw the universe for what it is. As the century began we knew little of the cosmos beyond our immediate stellar neighborhood. By the time it was three-quarters over we knew that galaxies exist, that stars are born and die, that the universe expands and was born in an eruption whose rumbling echo we still can hear. The age of the expanding universe has been determined and telescopes have plumbed to the edge of cosmic history. This rich fund of knowledge was acquired by a relatively small community of scientists working with modest resources over only a few decades. It is a dramatic story. Is it true?

Reports of scientific discovery, this book included, I'm afraid, tend to give a spurious impression of great progress recently attained. Their thesis is that humankind labored in ignorance for centuries until a few years ago, when the light of wisdom dawned. I think this tendency comes about because discoveries, by their nature, make good stories, while enduring bafflement does not; the storyteller concentrates on what has been learned and ignores what has resisted comprehension. In any event, it is a distortion. Now as before, the unknown dwarfs our little precincts of the known. Newton said, "To myself I seem to have been only like a boy playing on the seashore, and diverting myself in now and then finding a smoother pebble or a prettier shell than ordinary, whilst the great ocean of truth lay all undiscovered before me."

The history of science resembles less a smooth machine of progress than a rummage drawer of tools broken and discarded. In the late nineteenth century physics was widely thought of as a closing book. Gravitation had been accounted for by Newton, heat by Maxwell and Boltzmann, electromagnetism by Maxwell and Faraday, with such apparent success that students were cautioned to expect to make no breakthroughs in their careers, but to content themselves with the thought that physics henceforth would consist of merely refining existing "laws." Then in 1900 Max Planck of Berlin discovered the quantum principle, and science was transformed. Almost every "law" that had seemed so solid was altered, diminished or discarded. Alfred North Whitehead, born in 1861, said in the 1930s, "Nothing, absolutely nothing was left that had not been challenged, if not shaken; not a single major concept. This I consider to have been one of the supreme facts of my experience."

To imagine that cosmology is immune from similar revolution is to invite a shock.

The Big Bang cosmology—by which I mean not any single theory but rather all violent-genesis models built on general relativity—accounted for the expansion of the universe and the quasar cutoff point, and accurately predicted the discovery of the cosmic background radiation. Those were remarkable achievements and upon the basis of them the theory must be termed a success. But to say that a theory is successful is not the same thing as to say it is true. The cosmology of Claudius Ptolemy survived for 1,300 years, and, with various modifications, predicted, with reasonable accuracy, the motions of the planets in the sky all that time. Yet its central assumption, that the sun and planets circle Earth, was false. The sun-centered cosmologies of Copernicus and Kepler improved the accuracy with which theory fit observation; the accuracy was increased further with Newton's and, in turn, Einstein's theories of gravitation. All were human inventions and none should be confused with the phenomena they sought to explain. It seems unlikely that "gravity" existed in nature and that Newton "discovered" it, or that nature contained a "spacetime continuum" before Einstein discarded gravity and erected the continuum in its place. Our theories are not "laws" that nature "obeys." They imitate the sky; they did not fall from it. So it aids clear thinking to keep in mind that the Big Bang theory is not the universe, but only a preliminary, incomplete description of how, in a few very limited ways, the universe may behave.

What we do not know is an ocean; I will point out only a few recognizable inlets and estuaries, and those briefly. I cannot go farther, in part because beyond the light of our small knowledge, ignorance compounds itself—we do not know what we do not know. I think this is what Fred Hoyle had in mind when, at a conference where a speaker remarked that "there are questions we cannot answer," Hoyle whispered to the physicist Philip Morrison that in science the answers are not important, the questions are. The scientist who asks the right question reconnoiters a new patch of the unknown, and may, with luck, bring it within the constricted but expanding boundaries of the known. To learn new ways of asking questions, therefore, is a powerful thing.

Fortunately the recent history of science offers a tale concerning just that business of question-asking. It is the advent of quantum physics, the revolution Whitehead spoke of, which changed not only science but, to some degree, the foundations of science, its deep presuppositions.

By telling a little of this story, and giving a few examples of how it has affected modern cosmology, perhaps I can hint at something of the form of the frontiers of science.

Planck developed quantum theory in response to the "blackbody" problem that plagued experimenters. The experiment in question consisted of observing the radiation of heated gas in a neutral ("black") enclosure. By hooking up a spectrometer and temperature-measuring device, an experimenter could chart the intensity with which the gas radiated energy at various wavelengths. But when this experiment was performed, it gave results in conflict with the predictions of classical theory. Each gas tested yielded a curve—wavelength against temperature—that had a characteristic "blackbody" contour, with its peak clearly related to the temperature of the gas, but the shape of the curve and location of the peak were not those predicted by existing physics. Attempts to reconcile theory and observation failed. Ultimately Planck realized that only by breaking with the scientific tradition in which he had been raised could he account for the blackbody results. After months of what he called the most intense work of his career, he derived the formula since called "Planck's law." It has accurately predicted the behavior of radiated energy in laboratory jars, the sun and stars, and, apparently, the universe at large—the cosmic background radiation discovered by Penzias and Wilson conforms to a Planck curve.

Planck saw that atoms radiate energy not in a smooth continuum, as had been assumed, but in discrete packets he called "quanta" (after the Latin "quantus," for "how much"). Quantum theory holds that nature acts, in a sense, like a bank teller, who can pay out a penny or two pennies but nothing in between. "There is no way out," Planck told his students. "We have to become accustomed to the quantum theory, and we shall see that it will penetrate into more and more fields of our physics."

Nineteenth-century science viewed matter and energy as two different worlds, each with its own set of "laws." Matter was said to be composed of particles, energy of waves. Planck's theory violated this well-established distinction by portraying energy as composed of "particles," the quanta. Within a generation, the Parisian physicist Louis de Broglie showed that matter could be viewed as waves, just as energy could look like particles. Indeed, De Broglie said in his Nobel Prize address in 1929, "to describe the properties of matter, as well as those of light, we must employ waves and corpuscles simultaneously."

The general reaction among researchers was to say, well, fine, particle theory and wave theory can both be applied to matter or energy, but ultimately nature must be one or the other. Which is it *really?* Particles or waves? Much of the theorizing and experimentation of early twentieth-century physics revolved around this question.

Two schools emerged and briefly coexisted, representing the old wave-particle question in new clothing. Physicists who were intellectually more comfortable with a particle universe showed that the waves could be interpreted as probabilities: The ultimate reality was particles, in their view, and the waves reflected the probability of finding particles at each given point along, say, a ray of light. Einstein pioneered this position in an early paper explaining light as composed of quanta he called photons. Those favoring wave theory, notably the Viennese physicist Erwin Schrödinger, demonstrated with equal plausibility that the waves were what counted, that particles were an illusion. Each side delighted in inventing experiments that made the other's position seem dubious. Both sides were equally successful. Schrödinger, followed by Dirac, then demonstrated that the two schools were mathematically equivalent, that they amounted to different ways of saying the same thing.

But what was nature *really?* A swarm of particles? An ocean of waves?

The twenty-six-year-old physicist Werner Heisenberg, studying under Niels Bohr in Copenhagen, concluded that the paradox of waves and particles could be resolved only by taking the role of the observer, the physicist, into account. This step led Heisenberg to establish the uncertainty principle, and it is fair to say that science has not been the same since.

As Heisenberg recalled it, he was set on the path to the uncertainty principle by a remark Einstein made in a private talk the previous year. Einstein was disturbed by what he thought to be Heisenberg's almost ruthless refusal to concern himself with elements of the physical world other than those that could be observed. Specifically, Einstein was concerned about a paper by the young Heisenberg that failed to accept the notion that electrons orbit the nucleus of atoms. Heisenberg refused to talk about electron orbits because no electron had ever been observed in orbit, and in all likelihood none ever could be.

"But surely you don't believe," Einstein said, as Heisenberg remembered the conversation, "that none but observable magnitudes must go into a physical theory?"

"Isn't that precisely what you have done with relativity?" Heisenberg asked. "After all, you did stress the fact that it is impermissible to speak of absolute time, simply because absolute time cannot be observed; that only clock readings, be it in the moving reference system or the system at rest, are relevant to the determination of time."

"Possibly I did use this kind of reasoning," Einstein said, "but it is nonsense all the same. . . . It is quite wrong to try founding a theory on observable magnitudes alone. In reality the very opposite happens. It is the theory that decides what we can observe. . . ."

Late one night in February 1927, Heisenberg walked in Copenhagen's Faelled Park, thinking about Einstein's remark, "It is the theory that decides what we can observe." That certainly seemed to describe the matter of wave theory versus particle theory; one could see the subatomic world as waves or particles, whichever one looked for. What if the question of which was "real" could *never* be finally decided? What if the sort of things physicists observed in that tiny cosmos was determined less by "reality" than by the methods they used to observe? Heisenberg knew, for example, that one could not observe an electron in an atom without knocking it out of the atom: The radiation needed to "illuminate" the electron for observation—in practice, gamma rays were employed for this purpose—would knock it out of orbit. This had been the bone of contention with Einstein, who thought of the orbit as "real" even though an electron in orbit could not be observed. Heisenberg wondered whether this fact of the limitation of observation might be the issue. Perhaps we can never observe the "real" world on such a small scale. If we cannot, Heisenberg felt, then speculation about what that world was really like was not a matter for science.

Back in his flat, Heisenberg attempted to determine just how much uncertainty was built into the situation. If a physicist wants to know exactly where a "particle" is at a given moment, he must interfere with it—wallop it with gamma rays or some other radiation—to find out, and in the process he will change its behavior, knocking it onto a new path. The price of knowing its position is that he must relinquish knowledge of how it might have behaved if he had let it alone. If on the other hand he wants to know the particle's behavior in space and time, how fast it is going, in what direction—and it is this behavior of great numbers of "particles" that generates "waves"—he must wait and see; a track in a cloud chamber or an oscilloscope will tell him, after the fact, where the particle went. The price of that observation is that the physicist cannot know exactly where the particle was at a given moment. The trade-off is not technological but fundamental. The subatomic world Heisenberg visualized was built of little uncertainty "boxes" we macroscopic observers can *never* peer into. We can squeeze the box one way, to make things look like particles, or the other way, to see waves, according to how we set up each experiment, but we cannot reduce the area inside the box. When Heisenberg computed the "amount" of uncertainty inside the box, he found it defined by Planck's constant.

When Bohr returned to Copenhagen—he had been away on a skiing trip—and was presented with Heisenberg's new theory, he went to work extrapolating it. The result was his "principle of complementarity." He demonstrated that the uncertainty principle was implied by the laws of conservation of energy and momentum, considered part of the foundation of physical science. Wave and particle models were complementary aspects of the same reality. The "real" nature of things, it appeared to Heisenberg and Bohr and has appeared ever since, must remain forever unknown.

I have attempted this shamelessly oversimplified account of early quantum physics in order to suggest that something of deep importance was going on in science. Science was beginning to leave behind its need to visualize, and with it was discarding some other trappings of traditional science, including the doctrines of force and causation.

Scientific theories must be logical. They must be expressible in terms of mathematics, the most rigorous logical system known. The bones of all physical theories are numbers. But in practice, theories also have had to "make sense" in terms of ordinary experience. The scientist, looking at nature, draws on more of his human experience than just mathematics. Science is as full of metaphors, good and bad, as is poetry.

Because humans are visually oriented (so much so that the retina is generally regarded as part of the brain), models of nature typically have been visualizable, have made sense to the "mind's eye." It was this custom that broke down when physics penetrated to the subatomic world, where the "objects" being studied are not much larger than a wave of light, and cannot be seen at all. Heisenberg realized that for science to go on, this fact would have to be taken into account. The relationship of the scientist to the material observed would have to become an acknowledged part of science.

With the fall of visualization came the fall, or at least decline, of causation and force, two concepts that had long been part of science. The logical positivist Rudolph Carnap traced the two doctrines to a common origin. They arose, he said, "as a kind of projection of human experience into the world of nature.

"When a table is pushed, tension is felt in the muscles," Carnap wrote. "When something similar is observed in nature, such as one billiard ball striking another, it is easy to imagine that one ball is having an experience analogous to our experience of pushing the table. The striking ball is the agent. It *does* something to the other ball that makes it move. It is easy to see how men of primitive cultures could suppose that elements in nature were animated, as they themselves were, by souls that willed certain things to happen. This is especially understandable with respect to natural phenomena that cause great harm. A mountain would be blamed for causing a landslide. A tornado would be blamed for damaging a village."

With the advent of twentieth-century physics, both force and causation appear to have been expunged from at least part of science. Einstein did away with the "force" long thought necessary to keep the planets moving in their orbits. For centuries it appeared that something—perhaps God—must push the planets along in their paths. Newton explained planetary motions in terms of a force of universal gravitation, but apologized for doing so. In relativity, planets are said to follow their orbits because the orbits describe geodesics, paths of greatest efficiency, in space and time. No force holds them in orbit or pushes them along. In quantum physics causation fails because, given enormous swarms of matter/energy on a tiny scale, the experimenter can predict behavior of "particles" only in terms of probabilities. Einstein, who helped introduce the statistical approach to physics, felt it was only an expedient, and that a cause-and-effect explanation of the behavior of every particle could be worked out if one had sufficient equipment and patience. But Heisenberg showed that a complete cause-and-effect account of subatomic behavior could never be attained by us in the macroscopic world; we are prevented by the uncertainty principle from ever tracing the individual interactions of even a small group of subatomic "particles." Therefore—and this assertion is as much philosophy as science—it is said to be useless to talk about causation operating in a realm where we can never see it or examine it at work. We might as well accept that the probabilities are real.

The decline of determinism may have been made possible in the West by the decline of God's perceived role in the cosmos. For centuries God was assigned the role of prime causer ("unmoved mover," in Aristotle's phrase). Thomas Aquinas was able to construct an impressive proof of the existence of God by placing Him at the head of the cosmic causal chain. With the rise of technology, attention focused on the supposed machinery of nature, and God was increasingly relegated to the role of the creator who set the machinery in motion but now seldom had to look after it, like a successful businessman who need no longer show up regularly at the office. By the beginning of this century, when the machine age began its slow termination, God had withdrawn so far from the scene as to be unable to defend nature against a noncausal philosophy of science.

Although talk of causation survives in modern science, something of the scientific world view has been changed at the root by the quantum principle. Atoms are ubiquitous, and if the subatomic world is nondeterministic, the broader universe must present to us forever an element of chance. It cannot be the precise machine science once portrayed it to be. For many young physicists, the fall of causation was liberating; science had discarded a heavy piece of needless luggage and could proceed afresh. Reading Heisenberg or De Broglie, one gets the sense of adventure that the seventh-century Buddhist Fa-Tsang expressed when he wrote, "Now that we understand that causes are really not causes, any arising will be wonderful."

Einstein, however, did not share in this enthusiasm. He never accepted the contention that because the subatomic world must forever appear to us in probabilities, it really *is* built upon probabilities. Einstein had been one of the first to demand that the role of the observer be considered in physical theories, but he would not agree that limitations upon observation must be considered the same thing as limitations upon nature. "God does not play dice," he said.

Einstein became isolated from the quantum physics he had helped found. As De Broglie wrote in the early 1950s, "Theoretical physicists are at present divided into two apparently irreconcilable groups. On the one hand, Einstein and his followers are trying to develop general relativity theory, while by far the great majority of theorists, attracted by atomic problems, are trying to develop quantum physics quite independent of general relativity." De Broglie found it "incredible that the two great theories of contemporary physics, the theory of general relativity and that of quanta, should so utterly ignore each other." Max Born wrote of Einstein, "When out of his own work a synthesis of statistical and quantum principles emerged which seemed to be acceptable to almost all physicists, he kept himself aloof and skeptical. Many of us regard this as a tragedy—for him, as he gropes his way in loneliness, and for us who miss our leader and standard-bearer."

As he lived out his final years in a white frame house on Mercer Street in Princeton, Einstein found himself set apart—enshrined, really—both by his minority position in the philosophy of science and by the international celebrity that had befallen him. His kindly face and rumpled sweaters had become symbols of incomprehensible wisdom. "Because of a peculiar popularity which I have acquired, anything I do is likely to develop into a ridiculous comedy," he wrote an old friend, the Queen Mother of

Belgium. To Born he wrote sadly, "In our scientific expectation we have grown antipodes. You believe in God playing dice and I in perfect laws in the world of things existing as real objects, which I try to grasp in a wildly speculative way."

Einstein died April 18, 1955. A beautiful tribute he wrote for Max Planck, the founder of quantum physics, in 1932, might serve Einstein himself as well:

"Many kinds of men devote themselves to science," Einstein wrote, "and not all for the sake of science herself. There are some who come into her temple because it offers them the opportunity to display their particular talents. To this class of men science is a kind of sport in the practice of which they exult, just as an athlete exults in the exercise of his muscular prowess. There is another class of men who come into the temple to make an offering of their brain pulp in the hope of securing a profitable return. These men are scientists only by the chance of some circumstance which offered itself when making a choice of career. If the attending circumstances had been different, they might have become politicians or captains of business. Should an angel of God descend and drive from the temple of science all those who belong to the categories I have mentioned, I fear the temple would be nearly emptied. But a few worshippers would still remain—some from former times and some from ours. To these latter belongs our Planck. And that is why we love him."

The rift between Einstein and the quantum physicists, like a thousand other less olympian debates, reflected the fact that science is far from achieving a unified view of nature. The very large is assayed via relativity, the very small by quantum physics, the middle ground by a variety

of theories of more or less solid reputation. Four "forces"—
or "interactions" to use a more modern term—are known to
physics: gravity, electromagnetism, and two that bind
atoms together, the so-called "weak" and "strong" forces.
Almost nothing is known about why these interactions exist
or why they relate to one another as they do. Why is the
electromagnetic force so much more powerful—10^{39} times
as powerful—as gravitation? No one knows. There is as yet
no theory that encompasses all physical science and
explains, in Eddington's phrase, "the wide interrelatedness
of things."

In cosmology we see the incompleteness of science
magnified. The Big Bang theory is only a sketch, not a
painting. It fails to tell us how galaxies, stars and planets
formed: If the universe began as a homogeneous soup, why
did it not stay so forever? It has little to say about why
natural "laws" such as the constant of gravitation should
have been dealt out as they were and not some other way;
what, for instance, is the role of time in the natural world?
It is silent on so fundamental a question as what the
cosmos is made of. Atoms? Quanta? The hypothetical
building blocks called quarks?

In an effort to close some of the gaps, exotic new
cosmologies have been composed to harmonize with the
Big Bang theory, while seeking to deepen it.

Several theorists have proposed that the force of gravitation, assumed constant by Newton and Einstein alike,
might weaken as time passes and the universe expands.
Cosmic history on their terms would resemble a tennis
game played on a court that gets slowly larger, while the
ball turns lighter and the racquet strings loosen. What
justification is there for taking such an idea seriously?

One is Mach's principle. Mach, we recall, proposed that the inertia of any body was the product of its interaction with all the other matter in the cosmos. The young Einstein was struck by the beauty of this idea and also by the strange fact that the *inertial* mass of an object is exactly equal, within the limits of experimentation, to its *gravitational* mass. In other words, if you weigh something with a spring scale, then set it down on a block of ice or other friction-free surface and pull it sideways with the same scale, you will find that the amount of pull needed to get it moving is the same as its weight. Reasoning that this was unlikely to be coincidental, Einstein assumed nature was trying to tell him inertial mass and gravitational mass were in fact identical. This precept became part of the basis of general relativity, and was the reason Einstein believed he had incorporated Mach's principle into the theory. At the time he did not know that the universe is expanding—that the distribution of matter, which Mach saw as determining inertia, is thinning out. De Sitter and others subsequently demonstrated that relativity failed to incorporate Mach's principle. It seems at least possible that the failure stems from Einstein's assumption that gravitation remains constant, when instead it may diminish as the universe expands.

In 1937, at age thirty-five, the English physicist Paul Dirac attempted to find, in the relationship between constants of physics, numbers that might reveal something of the basic structure of the world. These were the "dimensionless numbers" that so fascinated Eddington. Dirac chose, as his fundamental unit of time, the interval required for light to cross the radius of a hydrogen atom. The age of the universe, expressed in these units, is slightly under 10^{40}. The coincidence with the ratio between the electromagnetic force and the force of gravitation, 10^{39}, suggested to Dirac that the age of the universe had something to do with the force of gravity. Alternately, gravity could be constant and the other numbers be changing, but the Machian associations of gravitation made Dirac favor gravitation as the flexible force.*

Gamow, Hoyle and Dicke as well as Dirac all proposed at one time or another that the constant of gravitation might be changing. The predicted rate of change is small, on the order of five or six parts in 10^{11} per year. By the mid-1970s experimental and observational technology was just approaching the sophistication needed to measure so subtle a rate of change. Physicists, astronomers and cosmologists waited to see whether the results would indicate that we live in a world where at least one of the laws previously thought immutable was flexible.

Another apparently changeless component of nature called into question is the passage of time.

Newton was obliged to make time grandly inflexible as space. "Absolute, true and mathematical time, of itself, and from its own nature, flows equably without relation to anything external," he wrote. This is time the grand monarch, commanding the sky that "impotently rolls as you or I," in the words of Omar Khayyám. But does it

---

* Something must remain constant, while something else changes, for the term "change" to have any meaning. To say *all* physical constants change at the same rate the universe expands is to say that none changes. The situation is reflected in two old philosophical puzzles. One asks, "How do we know but that the universe was formed *in toto* just last night, complete with all our supposed memories?" The other asks, "What if everything in the cosmos doubled in size overnight? How would we know anything had happened?" On examination, both questions can be seen to be empty by definition. In the first, the expression "last night" is meaningless because time must be transcended to accept the premise as plausible. In the second, space is transcended and so the term "double in size" has no meaning. Things can change in size only if something remains the same to measure the change by.

make sense to speak of time "without relation to anything external?"

Most of the world's religions and many of its philosophers, among them Spinoza, Kant, Hegel and Schopenhauer, denied that time is real. "To realize the unimportance of time," wrote Bertrand Russell, "is the gate of wisdom." At first sight this position may seem impractical, since we all act as if time were real; Kant was obsessively punctual, and had Spinoza's parents not fled the Spanish Inquisition *in time,* there would have been no Spinoza. The philosophers' point was not that there is no such thing as time but that there is no such *thing* as time; time, they held, is an attribute, like color or harmony, not a fundamental. Things generate time, not the reverse. Lucretius put it this way: "Time does not exist by itself, but from occurrences a feeling arises by which we distinguish what has taken place in the past, what now exists, and what may follow hereafter. We must admit that no one perceives time in itself apart from the motion or immobility of matter." Leibniz wrote, "Instants, considered without the things, are nothing at all. . . ."

Einstein called time a dimension, but if so it is a very strange dimension. We are permitted to move forward or backward in the three dimensions of space, but time flows one way only, toward the future. Eddington called this phenomenon "the arrow of time." What is the reason for it?

Taking a clue from Lucretius and Leibniz—that time cannot be divorced from things—imagine a Big Bang universe "before" expansion began. All matter and energy are compressed to high density. The universe is chaotic, matter and energy swarming incoherently. There are no regularly recurring motions by which to measure the passage of time, and no observers who could note them if there were. It seems impossible to avoid the conclusion that there can be no time in such a universe. In the classic Big Bang model, this means we cannot inquire about the cosmos prior to the beginning of expansion, cannot know whether the compressed universe existed for an instant or a billion years before expansion began—indeed the question is rendered meaningless—and it was just that feature of the theory that led some cosmologists, Hoyle among them, to reject it.

Now imagine that expansion has begun. In this case we can see a coherent trend in the behavior of the particles: They are thinning out. Two imaginary photographs taken in an expanding universe would show the cosmos at differing densities. In one, the atoms (or galaxies, depending upon at what epoch the ethereal photographer trips the shutter) are closer together than in the other. Comparing the two, we say, "Ah ha! Time is passing. One of these photographs was taken *after* the other." As Plato said, "Time and the heavens came into being at the same instant."

Notice, though, that while the two photographs will, in themselves, imply that time is passing, they will not tell us in which *direction* it is passing. A motion picture taken before expansion began and another taken after expansion began (the critical reader is asked to pardon the use of "before" and "after") will have in common that either can be run "forward" or "backward" and make equal sense. The preexpansion film will show a chaotic jumble in either case. The postexpansion film will show galaxies either thinning out or coming together, depending upon which way we run it. If we know the universe is "expanding," we can tell the projectionist which is the right way to roll the film, but to say the universe is "expanding" means we have already intuited the arrow of time: expansion *means* that spatial intervals increase as time passes. This intuition comes naturally to us because we live in the universe, we progress from birth toward death, and we are accustomed to the feeling that time flows in one direction only.

So are we back where we started?

Almost. We can see that the sensation of time may be generated by the cosmos. If the cosmos is expanding, it certainly is possible that expansion is responsible for the arrow of time pointing in only one direction. But *how* is it responsible? What does the wide universe do to keep our clocks ticking and hearts beating, to pace us through our lives just as if time were real, if our considered reflection tells us time is not real? Until we identify a substructure linking us with the universe at large, the nature of time is likely to remain a mystery.

Several cosmologies envision the arrow of time reversing. Most are closed-universe models where the present expansion is succeeded by contraction into a fresh fireball, which in turn produces a new universe where time runs backward relative to the one before. Our future is seen as the past of the "next" expanding universe. For time reversal to have meaning, of course, enough information must survive through the fireball so that a trace of the last (and future) cosmos remains. Some have proposed that the laws of thermodynamics may contain information of that sort, but searching for it probably lies beyond the scope of present research. A unique and difficult cosmology constructed by Thomas Gold suggests that the universe oscillates and time reverses, not in the fireball, but at the point of maximum expansion. When contraction begins, time commences to flow backward. The local effects are at first subtle, later profound—something like those Mach described when he referred to "the famous lobster chained to the bottom of the Lake of Mobrin, whose direful mission, if ever liberated, the poet [August] Kopisch humorously describes as that of a reversal of all the events of the world; the rafters of houses become trees again, cows calves, honey flowers, chickens eggs, and the poet's own poem flows back into his inkstand."

In 1973 Richard Gott proposed an intriguing cosmology incorporating time reversal and antimatter. The existence of antimatter had been hypothesized by Dirac in 1928 on the basis of both relativity and quantum physics, which, he saw, required that each category of known subatomic particles ought to be matched by a corresponding kind of particle with equal mass but opposite charge. If a particle met an antiparticle, Dirac's calculations indicated, the two would annihilate each other in a flash of energy. The prediction seemed so unlikely that Dirac himself called it a "blemish" on his work, but in 1932 a physicist at Cal Tech who had never read Dirac's theory discovered, in a cloud chamber, a bit of antimatter. It developed that a complete family of antiparticles does exist, as Dirac had envisioned. Very little antimatter exists on earth; if it did we would live in a maelstrom. A question for cosmology therefore is whether there are stars and galaxies made of antimatter (present-day observations cannot say, because light from antiatoms looks just like light from atoms) or, if not, why the universe should have been made asymmetrically, favoring matter over antimatter. Gott's contribution to the problem was to picture the Big Bang as generating not one but three universes. The first, Region I, is where we live; time flows "forward" (that is, in the direction we are accustomed to) and matter predominates over antimatter.

Region II emerged from the fireball traveling in reverse time; it shares with ours our common genesis, but since our arrows of time point in opposite directions, the two universes have moved apart in time ever since, and one cannot observe the other. Region III is a universe of tachyons—theoretical particles that travel faster than light.* The tachyon universe has outrun both other universes since the beginning—they are both limited to the speed of light—and it can be observed from neither. Gott's Region II lies permanently in our past, Region III permanently in our future. Both can be confronted only in another fireball. An ornament to Gott's cosmology is provided by the physicist Richard Feynman, who discovered that antiparticles can be viewed as ordinary particles moving in reverse time. The stray antiparticles we find in our universe could be said to resemble souvenirs of the stuff that makes up the time-reversed universe of Region II, where antimatter physicists study traces of what we consider ordinary matter and, perhaps, wonder why there isn't more of it where they are.

Among the most provocative of the new cosmological theories is John Wheeler's "superspace." Wheeler, a Princeton University physicist who studied under Bohr and taught Dicke, often thought about the rift between relativity and quantum physics, wondering where clues might be found to their reconciliation. A possible answer came to him when he began working on the theory of black holes. Wheeler, who coined the term "black hole," thought of them as a meeting ground between general relativity, which predicted their existence, and quantum physics,

---

* Relativity theory forbids any particle to be *acccelerated* to the speed of light, but permits tachyons because they have *always* been traveling faster than light and will never slow down to it; that tachyons exist has not been established.

which governed the world of the very small. What did quantum theory imply about the nature of space on an extremely small scale?

Applying both relativity and quantum physics, Wheeler found that there was a unit of size such that, if a single quantum were smaller than that size, the quantum would act like a tiny black hole. The gravitation of that quantum, though of course minuscule, would be strong enough to wrap spacetime around the quantum and remove it from the observable universe. Since nothing smaller than this could possibly be observed, space itself may be considered quantized. The size of the quanta, known as the Planck-Wheeler length, is 10^{-33} centimeter.

Wheeler put it this way: "From an airplane six miles high, the ocean looks smooth. Down at sea level, in a life raft, however, we see that waves are breaking, and the surface is highly irregular; what's more, instead of being merely irregular, there are droplets breaking loose. Now space, too, looks smooth at the scale of everyday life, smooth at the scale of atomic structure, and smooth at the scale of nuclear structure. But when one gets down to the scale of distances 20 powers of ten smaller than the scale of nuclear structure, then one predicts that space is foamlike.

"At first this theoretical prediction seems like the most esoteric thing in the world. But recall: there was a time when we thought of a piece of wood as solid; then we learned to think that it's 99.9 percent empty space and that it derives its solidity from electrons whirling around. And now we dare to speak about the possibility that the electrons themselves are, in some sense difficult to be specific about, also made of empty space; or, rather, space that is changing, dynamic, altering from moment to moment."

Owing to the schism between relativity and quantum theory, most modern cosmology has pictured the universe as a relativistic arena all the inhabitants of which —the matter and energy population—were governed not by relativity but by quantum physics. Wheeler's quantizing of space represented an attempt, without abandoning relativity, to bring the arena of space under quantum theory as well. "No principle that we know of in all of physics has the same universal power as the quantum principle," he said. "The more we pursue it, the more it looks as if it is *the* number one principle, and that everything else is, in some way we don't yet understand, derived from it."

The uncertainty principle is broadened in Wheeler's theory to encompass space as well as matter and energy. The cosmological geometry of space is seen as only a *probable* geometry, the sum of the uncertainties of all the space quanta in the universe. Determinism is removed from cosmology; the outward march of the galaxies is not absolutely determined by the force of the Big Bang, as in most theories, but is to a degree uncertain, and whether the universe will expand forever depends upon quantum uncertainties in the fireball.

Superspace is a hyperdimensional grid Wheeler constructed in which to plot probable pasts and futures of his quantized-space cosmos. It can be represented as a triangular dish. Within it can be traced the evolution of various possible spacetime geometries, various cosmoses. The universe of the present is presented as a point, its evolution in time as a line. The walls of the dish represent gravitational collapse. Our cosmic history describes a line departing from one wall and proceeding toward the center of the dish; the line is fuzzed by uncertainty but is hump-shaped in the middle, where the average of all the uncertainties proofs out to approximate the Einsteinian, Newtonian cosmos we see. An oscillating universe wanders in the dish for a while, skids toward one wall, rebounds off it (the Big Bang), and starts a new path of expansion by vote of chance.

The geometry of our universe is only one of many we might have been dealt, in Wheeler's view. There is room in superspace for an infinite number of geometries. "Imagine a gigantic auto body part shop that has parts for every automobile ever made," Wheeler said. "Stacked up on the big field are auto fenders of the earliest Ford over in a corner, then as the Ford evolved, the shape of the fender began to change a little, so those altered fenders are stacked a little further along in the field. The man who has the miserable job of locating all the parts has had the wit to place them so that fenders of nearly the same shape lie at nearly the same position, so they can be found easily. This big field would be superspace, and the automobile fenders the various geometries of space."

The cosmic fireball might have dealt a wildly different universe than the one we inhabit, with an alien geometry and different physical laws; many of the universes possible in superspace never form stars, planets, or even atoms and molecules. This brings us, by a variant route, to a point we have considered before, that we owe our small existence to that of the broad universe. It suggests, in Wheeler's words, "that there exists a degree of harmony between us and our surroundings that we never realized before. In the past, we looked at our surroundings as if there could be no other, something with which we just had to get along. If this new view is correct, our surroundings are very special and tuned to us, like a plant to its flower: This cycle of the universe like the plant, and we like the flower."

Through all science runs that refrain, that nature is intelligible to us because we in some sense belong to it. The scientist, wrote Einstein, "is astonished to notice how sublime order emerges from what appeared to be chaos." Carl Friedrich von Weizsäcker told his friend Werner Heisenberg, "All our thinking about nature must necessarily move in circles or spirals; for we can only understand nature if we think about her, and we can only think because our brain is built in accordance with nature's laws." James Jeans wondered of the galaxies, "Do their colossal incomprehending masses come nearer to representing the main ultimate reality of the universe, or do we? Are we merely part of the same picture as they, or is it possible that we are part of the artist? Are they perchance only a dream, while we are brain-cells in the mind of the dreamer?" "Ultimately," wrote Mach, "all must form one whole."

To say that nature is comprehensible to us, that science is not deluding itself, is an assertion of faith—"reason is one of the articles of faith," said Eddington—but there is nothing wrong in that. After all, we are part of the universe. And the faith of science—that the seamless wave of nature will reveal itself to our reasoned inquiry—that faith is part of the universe too.

GLOSSARY

Astronomical nomenclature can be confusing, as when names given in the first flush of discovery later prove inappropriate. For example, galaxies are still sometimes called "spiral nebulae," though they are not nebulae at all; the original name persists despite the fact that their real nature has been understood for generations. Quasi-stellar objects, or quasars, were so named because they looked something like stars. It has since become evident that they are not stars, and instead may be the cores of forming galaxies; but the name has gained currency and it appears to be too late to change it.

The brief definitions that follow are offered in the hope that they will help in understanding the more technical or idiosyncratic terms that appear in this book.

ABSOLUTE ZERO. The temperature at which a gas has no thermal energy. It equals Minus 459.69 degrees Fahrenheit, or zero degrees Kelvin. (The Kelvin scale is also known as the Absolute scale.) According to measurements of the *cosmic background radiation,* the universe has a temperature of about 2.7 degrees (Kelvin) above absolute zero; if the *Big Bang theory* is correct, this represents the state to which the cosmos, once as hot as billions of degrees or more, has now cooled.

ANTIMATTER. Matter made of particles the same as those of ordinary matter except that their charge is reversed. The proton, for example, has a positive charge, while the antiproton has a negative charge. Note that the term "anti" merely reflects our predilection for the sort of matter we are made of, and ought not to be taken as suggesting that nature is in any way biased against antimatter, which may, for all we know, be as abundant in the cosmos as ordinary matter.

BIG BANG THEORY. The hypothesis that *expansion of the universe* commenced when the universe—all space, time, matter and energy—was concentrated to high density; expansion can be said to have begun when the density of the cosmos began to drop, as it continues to do today. The universe is not supposed to have expanded into space; all space was then, as it is now, inside the universe. If we think of time, the fourth dimension, as having just begun then, we can imagine a three-dimensional universe wrapped up rather tightly in the fourth dimension. The *Big Bang* is an umbrella term for a variety of cosmologies, all having in common that they propose that expansion began violently from a high density cosmos.

BILLION. The billion employed in this book equals 1,000,-000,000, or one thousand million.

BINARY STAR. A system of two stars orbiting a common center of gravity. Such systems are very common; indeed, our solar system came close to forming one. If the planet Jupiter were 50 to 100 times more massive than it is, thermonuclear processes would take place in its core, it would be a small star, and the sun would belong to a binary system.

BLACKBODY. An idealized object that absorbs all radiated energy striking it, and so appears black. The concept is useful in physics, where experiments are performed employing an enclosure with blackened interior to approximate a perfect blackbody. The universe as a whole should in theory be a blackbody; for this reason, observations indicating that the *cosmic background radiation* has the characteristics of blackbody radiation are taken as evidence that the background radiation is universal, and probably results from energy released in the *Big Bang*. See *Planck Curve*.

BLACK HOLE. A theoretical object so compressed that its gravity closes spacetime around itself, so that no matter or energy can emerge from it. A black hole can be regarded as closed off from our universe. Black holes, if they exist, may result from the gravitational collapse of stars, as in the aftermath of a *supernova*.

C. Designates radio sources in the Cambridge catalogues of cosmic radio sources. Several editions have appeared—3C for the third catalogue, 4C for the fourth and so on.

CEPHEID VARIABLE STAR. A giant pulsating star whose period of variation is related to its *absolute magnitude*. This quality makes Cepheid variables useful to astronomers as distance indicators. The astronomer observes a Cepheid over a period of months and notes how long it takes to vary in brightness. Its absolute magnitude can then be inferred from its period. Comparing the absolute magnitude with the *apparent magnitude* of the Cepheid in the sky, the astronomer can then say how far away the star is. This method has been invaluable in finding the distances of globular clusters and other galaxies. See *Inverse Square Law*.

COSMIC BACKGROUND RADIATION. Residual energy pervading the entire cosmos, left over from the violent fireball believed to have begun the expansion of the universe. Its existence was proposed by George Gamow as a consequence of his *Big Bang theory*. If the universe began in a Big Bang, the energy radiated by that event should still be present everywhere, he reasoned, though at a level much reduced because it has thinned out with the expansion of the universe. According to the laws of quantum physics, such radiation should conform to a trace of energy against wavelength called the *blackbody* curve. Arno Penzias and Robert Wilson observed what appears to be the cosmic background radiation in 1965; its energy level (often called "temperature") is close to Gamow's prediction and it describes a blackbody curve.

COSMIC RAYS. See *Electromagnetic Radiation.*

COSMOGONY. The study of the origin of the universe, its galaxies, stars and planets. See *Cosmology.*

COSMOLOGICAL CONSTANT. A term introduced into the general relativity equations by Einstein, who later discarded it. It represented an attempt by Einstein to construct a static model of the universe, against the implication of relativity that the cosmos was dynamic instead. The discovery of the *expansion of the universe* eliminated the need for the cosmological constant, though it still shows up occasionally in *cosmology,* usually as only a mathematical convenience.

COSMOLOGICAL PRINCIPLE. The assumption that the cosmos at large is generally the same at all locations—that is, all observers see galaxies in all directions, find them to be rushing apart with the *expansion of the universe,* and so forth. The principle serves to affirm that the same physical laws operate throughout the universe—without which assumption it would be very difficult to create cosmological theories—and it bars special pleading, such as arguing that the Milky Way is the center of the universe. In short, our view of the cosmos is said to be typical. See *Perfect Cosmological Principle.*

COSMOLOGY. The science concerned with the nature of the physical universe as a whole. Scientific cosmology involves elements of astronomy, physics, mathematics and philosophy, and cosmological theories normally are created by scientists from one or more of those fields of study; there are few if any full-time cosmologists. To put it another way, a theorist should have done solid work in a less speculative scientific area if he or she expects to publish in cosmology and be taken seriously. Cosmological theories are expected to conform to the laws of physics (or at least most of them), to be consistent with the astronomical evidence, and to seem at least marginally reasonable. Philosophical considerations are always present, though more conspicuous in some theories than in others.

DECELERATION PARAMETER. The rate at which the expansion of the universe is slowing down. If, as some observations indicate, the deceleration parameter approximates or equals zero, expansion will go on forever. See *Expansion of the Universe.*

DOPPLER SHIFT. A displacement in the wavelength of light or other radiation caused by the motion of an object, such as a star or galaxy, relative to us. If a given galaxy and ours are moving away from each other, the spectral lines in light from that galaxy will be displaced toward the red end of the spectrum. See *Redshift.*

DWARF STAR. A star of small to medium size and *luminosity.* See *White Dwarf.*

ELECTROMAGNETIC RADIATION. Energy propagated by means of electric and magnetic fields. Its velocity in a vacuum is that of light, and, indeed, light is one form of electromagnetic radiation. Other wavelengths of electromagnetic radiation are described by a variety of names. A tour of the electromagnetic *spectrum*, from long wavelengths to short wavelengths, reads like this: Longwave radio, shortwave radio, VHF television, UHF television, radar, microwaves, infrared light, visible light, ultraviolet light, X-rays, gamma rays, cosmic rays.

EXPANSION OF THE UNIVERSE. The motion of galaxies away from one another at velocities proportional to their separation. Another way to put it is to say that the average density of the universe is decreasing, as, according to the *Big Bang theory*, it has been since genesis. The motion of galaxies is inferred from the redshift in their light, for which the only satisfactory explanation known is that they are rushing apart. Expansion pertains not among closely neighboring galaxies—clusters of galaxies like our *Local Group* are bound together gravitationally—but between clusters. See *Doppler Shift, Hubble's Law, Redshift.*

EXPONENTIAL NOTATION. In order to deal conveniently with the large numbers that pervade astronomy, exponential notation in the power of 10 is used. The rule is that the exponent—the small number to the upper right of the 10— signals how many places to move the decimal point to the right or left. Positive exponents move the decimal point to the right: One million, or 1,000,000, becomes 10^6, while an American *billion*—the billion used in this book, 1,000,000,-000—is noted as 10^9. Negative exponents move the decimal point to the left: One one-thousandth, or .001, is 10^{-3}. Precise numbers are expressed by multiplication; the mean distance from Earth to the moon in miles, 240,000, can be written 240×10^3, or 24×10^4. The approximate radius of the universe in light-years, 15,000,000,000, becomes 15×10^9.

EXTRAGALACTIC NEBULA. See *Nebula.*

FUSION. See *Nucleus.*

GALACTIC NEBULA. See *Nebula.*

GALAXY. An enormous system of stars, dust and gas. Some are flattened spirals, others almost spherical, others irregular splashes of stars. Our sun belongs to the *Milky Way* galaxy, a large spiral shaped something like a plate or a pinwheel. Galaxies are grouped in clusters, which in turn appear to be organized into vast superclusters. The degree to which this hierarchy pertains in the cosmos—are there clusters of superclusters?—is unknown. The number of galaxies in the universe is estimated at 100 billion, comparable to the number of stars in our galaxy. See *Milky Way.*

GAMMA RAYS. See *Electromagnetic Radiation.*

GLOBULAR CLUSTER. See *Star Cluster, Globular.*

HERTZSPRUNG-RUSSELL DIAGRAM. A plot of the absolute magnitude of stars against their temperature or color. A distinctive pattern emerges, a kind of forking path called the *main sequence.* By diagramming stars that lie both on and off the main sequence, astrophysicists have found valuable clues to stellar evolution and the origins of star clusters. Analysis of the Hertzsprung-Russell diagram helped Walter Baade discover that stars belong to distinct generations, or populations.

HUBBLE'S LAW. The rule that light from distant galaxies is shifted toward the red to a degree directly relating to how far away each is. Also known as the velocity-distance relationship, because velocity is believed to be the means by which the redshifts are generated. See *Doppler Shift, Expansion of the Universe, Redshift.*

INVERSE SQUARE LAW. The rule that the apparent brightness of an object decreases in proportion to the square of its distance. A star ten light-years away looks one quarter as bright as the same star five light-years away.

LIGHT. See *Electromagnetic Radiation*.

LIGHT-YEAR. The distance light travels in one year. A light-year equals 5.88×10^{12} miles, or 5,880 billion miles. See *Exponential Notation*.

LOCAL GROUP. A cluster of galaxies that includes the *Milky Way* galaxy, the large spiral galaxy in Andromeda, and perhaps two dozen other systems. The number of galaxies in the Local Group is not known exactly, because the *Milky Way* blocks our view of part of the cosmos.

LOOKBACK TIME. The amount of time light from a remote object, such as a galaxy or quasar, took to reach Earth. The phrase reflects the fact that when we observe across vast distances of space we are also looking back in time; a galaxy 100 million light-years distant, for instance, appears as it was 100 million years ago.

LUMINOSITY. The amount of energy a star or other body radiates into space. See *Magnitude*.

M. Designates an object in the *Messier Catalogue*.

MACH'S PRINCIPLE. Broadly stated, an attempt, by Ernst Mach, to express what Arthur Stanley Eddington called "the wide interrelatedness of things." The "weak" form of Mach's principle holds that the motion of one body requires the existence of other bodies against which the motion can be measured; it is meaningless to speak of the "motion" of a single object relative to nothing. The "strong" form of the principle extends this view to argue that not only motion but the inertia of objects depends upon the interrelation of all matter in the cosmos. In both forms Mach's principle influenced a number of physicists, particularly Einstein. However, it is a philosophical assertion, not a scientific one. No mathematical formulation of Mach's principle has appeared, and no one has been able to determine how his purported interrelations of matter propagate themselves.

MAGELLANIC CLOUDS. A pair of irregular galaxies that are satellites of our *Milky Way* galaxy. They can be seen in the skies of Earth's southern hemisphere, where they resemble detached pieces of the *Milky Way.*

MAGNITUDE. The measure of brightness of astronomical objects. The dimmest star visible to the unaided eye is apparent magnitude 6, while a bright star like Aldebaran is magnitude 1. Still more brilliant objects are designated by 0 or by negative numbers, so that Sirius, the brightest star in Earth's skies, is magnitude –1.6, the full moon –12 and the sun –27. See *Magnitude, Absolute,* and *Magnitude, Apparent.*

MAGNITUDE, ABSOLUTE. Intrinsic brightness. Astronomers define it as the *apparent magnitude* a star or other body would have if viewed from a distance of 3.26 light-years. The absolute magnitude of the sun is about 5. Most stars have absolute magnitudes between 0 and 15, though dwarfs as dim as magnitude 20 have been observed, as have a few dazzling giants as bright as –10. See *Magnitude.*

MAGNITUDE, APPARENT. How bright a star or other astronomical object appears in the sky. Apparent magnitude is a function of each star's *absolute magnitude,* how far away it is, and whether interstellar dust or gas dims its light in transit to us. See *Inverse Square Law.*

MAIN SEQUENCE. See *Hertzsprung-Russell Diagram.*

MESSIER CATALOGUE. A list, compiled by the French astronomer Charles Messier, of star clusters, nebulae and other objects in the sky of interest to observers with small telescopes. Some of the Messier objects belong to our galaxy. Others, among them M31 in Andromeda, are distant galaxies in their own right.

MICROWAVE. See *Electromagnetic Radiation.*

MILKY WAY. A glowing band of soft light in the skies of Earth. It represents the plane of our galaxy seen from within. Because the galaxy is flattened in shape, its stars, dust and gas are much more concentrated along the plane, giving rise to the Milky Way's appearance as a sort of river in the sky. All the individual stars we see with the naked eye, however, both in the Milky Way and elsewhere, belong to our galaxy. See *Milky Way Galaxy.*

MILKY WAY GALAXY. The galaxy to which our sun belongs. It is a flattened spiral perhaps 100,000 light-years in diameter; the sun lies about two-thirds of the way out from the center toward one edge. See *Galaxy.*

NGC. Betokens objects listed in the *New General Catalogue,* a list of star clusters, nebulae and galaxies published in 1888.

NEBULA. In the broad sense, almost any astronomical object that looks indistinct in a telescope. *Gaseous nebulae* are clouds of gas within our galaxy; *galactic nebulae* are clouds of gas or dust, some illuminated by nearby stars, others dark and visible only in silhouette; all lie within our galaxy. The so-called *extragalactic nebulae,* or *spiral nebulae,* are other galaxies. The misnomer dates from times before telescopes were large enough to determine the difference between other galaxies and clouds in our own galaxy, so that the Andromeda Galaxy, M31, for example, is still sometimes called the Andromeda Nebula.

NEUTRON STAR. An extremely dense object formed from the collapse of a dying star. Pulsars are so named because their gravitational force is great enough to overcome the structure of their atoms, crushing them down to nuclei. See *Nucleus.*

NOVA. Explosion of a star. See *Supernova.*

NUCLEUS. In atomic physics, the central part of an atom, around which swarm electrons. Nuclear *fusion,* the interchange of particles from atomic nuclei in which a small part of their mass is converted into energy, is the mechanism that makes the stars shine.

PARALLAX. An apparent shift in the position of a star in the sky, caused by a change in the location of its observer. If, for example, a relatively nearby star is photographed from Earth in June and again in December—when Earth will have traveled halfway around the sun—its position against more distant background stars will appear to have shifted. What has changed, of course, is our vantage point. (The effect can be illustrated by holding one finger out at arm's length and sighting on it with first one eye, then the other. One eye represents the Earth's position in June, the other in December, and the shift of the finger against background objects is the parallax.) Parallax is used to determine the distances of stars, but because even the nearest stars are quite far away, parallax displacements generally are minor, and the method is accurate out to only a few hundred light-years.

PLANCK CURVE. A tracing of wavelength against intensity

of radiation formulated by Max Planck to account for the behavior of *blackbody* radiation. Planck's *quantum* theory correctly predicted the behavior of blackbody radiation where previous theories could not. See *Quantum.*

PLANCK'S CONSTANT. The key expression in Planck's quantum theory of radiation. See *Quantum, Blackbody.*

PLANET. A body that shines by reflected light and orbits a star. The sun has nine major planets.

PULSAR. A rapidly spinning *neutron star* that gives off pulses of radio noise.

QUANTUM. A fundamental unit of energy hypothesized by Max Planck to explain the behavior of gas in *blackbody* experiments. The quantum principle portrays energy as released in discrete packets, rather than in a continuous stream. It has been applied, with considerable success, to matter and space as well.

QUASAR. A compact, bright object which typically is found at great distance, according to measurements of the *redshift* of quasar light. Several theories hold that quasars are galaxies in early stages of formation.

RADIO ASTRONOMY. Observation of the heavens in radio wavelengths. Many astronomical objects, from the sun and the planet Jupiter to giant galaxies, emit radio naturally at various wavelengths, and so can be observed with radio receivers, called *radio telescopes.*

RADIO TELESCOPE. A device for receiving radio radiation from galaxies, clouds of interstellar gas and other natural cosmic sources. (No artificially generated radio signals have yet been received from space, except for those broadcast by space vehicles launched from Earth.) Because radio waves are much longer than light waves, radio telescopes generally must be constructed on a larger scale than optical telescopes: Dish antennas of 100 feet or more in diameter are numerous, and the world's largest radio telescope, at Arecibo, Puerto Rico, is 1,000 feet wide.

RADIO WAVES. See *Electromagnetic Radiation.*

REDSHIFT. Displacement of the light received from a star, galaxy or other object toward the lower-frequency end of the *spectrum.* (Displacement toward the *higher*-frequency end of the spectrum is called blueshift.) Redshifts are measured by identifying spectral lines of elements prevalent in the object (such as hydrogen) and comparing their position with a standard spectrum obtained by burning a small sample of the same element in the laboratory or observatory. Redshifts may be produced in several ways, one of which is *Doppler Shift:* The light from, say, a distant galaxy is displaced toward the red because our own and the other galaxy are rushing away from each other, just as a train whistle sounds lower in tone if the train is speeding away. Galaxies do display redshifts, which increase in direct relation to the distance of the galaxy in question; this is primary evidence of the *Expansion of the Universe.* See *Hubble's Law.*

REFLECTING TELESCOPE. A telescope that employs a curved mirror, rather than a lens, to gather light. Most large astronomical telescopes are reflectors.

REFRACTING TELESCOPE. A telescope using a lens to gather light. The classic "spyglass" is a refractor.

RELATIVITY, GENERAL. Einstein's theory of gravitation. See *Relativity, Special,* and *Spacetime.*

RELATIVITY, SPECIAL. Einstein's theory of motion and of the propagation of electromagnetic radiation. The theory can be said to be based upon the idea that the universe harbors no fixed points of reference, no absolute space or time. Measurements of physical behavior must be made from a given observer's frame of reference, and no one frame of reference has priority over another.

SPACETIME. A construct invoked in Einstein's theory of *general relativity*. Events in the physical world are viewed as taking place in a four-dimensional continuum, three dimensions of space and one of time. The concept has won widespread acceptance due to relativity theory's virtually flawless record of accurately predicting and accounting for physical phenomena.

SPECTROGRAPH. A spectroscope with a device attached, usually a camera, to record the spectrum for later study. See *Spectrum*.

SPECTROHELIOGRAPH, SPECTROHELIOSCOPE. Instruments for viewing and photographing the sun and its atmosphere in detail. This is made possible by restricting the light observed to a narrow band of the spectrum.

SPECTROSCOPE. An instrument for observing light arrayed in its various wavelengths. See *Spectrum*.

SPECTROSCOPIC BINARY. A *double star* in which the two stars are too close together to be seen as individuals by telescopes on Earth. Spectroscopic analysis of light from the system reveals that it is a binary system, hence the name.

SPECTRUM. Radiated energy broken down into its component wavelengths. In optical astronomy, the energy involved is light. A *spectroscope* spreads the light out in an array according to wavelength, with violet, the shortest wavelength of visible light, at one end, and red, the longest, at the other. A series of lines along the spectrum, generated by a narrow slit in the instrument, signals the presence of various elements under various conditions. These include dark *absorption lines* and bright *emission lines;* by studying them a specialist often can tell a great deal about what a star, planet or galaxy is made of and what conditions prevail there. In studying distant galaxies, measuring the degree to which the spectral lines are displaced toward the red is important, because this is taken as an indication of how far away the galaxies are in the expanding universe. See *Doppler Shift, Hubble's Law, Redshift.*

SPIRAL NEBULA. A spiral galaxy. See *Galaxy.*

STAR. A self-luminous body of gas sufficiently compressed for thermonuclear processes to operate at the core.

STAR CLUSTER, GLOBULAR. An assembly of stars gathered within a spherical volume of space. The centers of globular clusters are so congested that several stars may be found within a single cubic light-year. Globular cluster stars are generally very old, and many of the clusters are found above and below the planes of spiral galaxies such as our Milky Way. Astrophysicists hypothesize that the clusters formed early in the history of galaxies, when the spirals may have been more nearly spherical in shape and had not yet taken on the flattened form we observe today.

STAR CLUSTER, OPEN. A loose group of stars. Presumably they are generated when a number of stars all form at about the same time out of a common cloud of dust and gas.

STEADY STATE THEORY. The theory that the universe is infinite in space and time, and has always been more or less as it is now. To allow for the *expansion of the universe,* Steady State theorists are obliged to hypothesize that matter is continuously created to fill in gaps left as the galaxies rush apart, so that the density of the universe at large may remain the same.

SUPERNOVA. An exceptionally violent explosion of a star. The detonating star may shine hundreds of millions of times its normal brightness for several days before fading. Supernovae are believed to be the means by which certain heavy elements, such as gold and uranium, were created.

TWENTY-ONE-CENTIMETER RADIATION. Radio emissions produced by hydrogen atoms in space. Since most of the matter in interstellar space is hydrogen, study of this natural radio noise has been invaluable in mapping our galaxy and others. See *Radio Astronomy*.

UNCERTAINTY PRINCIPLE. An axiom holding that the location and motion of a subatomic particle cannot both be known with exact accuracy. If the experimenter seeks to know the exact motion of the particle, he must sacrifice knowledge of its precise position, and vice versa. Therefore a full account of particle behavior can be known only in terms of probabilities.

VARIABLE STAR. A star that changes in *luminosity* regularly. Some variables have periods of less than a day, others weeks, months or even years. Some change dramatically; others so little as to be almost undetectable. See *Cepheid Variable Stars*.

WHITE DWARF. A dense, compact, glowing star believed to be formed from a previously healthy star that exhausted the fuel at its core and collapsed.

X-RAYS. See *Electromagnetic Radiation*.

SELECTED
BIBLIOGRAPHY

HISTORY, BIOGRAPHY, MEMOIRS

THE MOST EXTENSIVE ACCOUNT OF THE DEVELOPMENT OF contemporary astronomy is the highly readable, though semitechnical, *Astronomy of the Twentieth Century,* by Otto Struve and Velta Zebergs (Macmillan, 1962). For a broader history see Fred Hoyle's *Astronomy* (Doubleday, 1962), Arthur Koestler's *The Sleepwalkers* (Macmillan, 1959), or G. de Vaucouleurs' *Discovery of the Universe* (Macmillan, 1957). Charles Whitney relates how the nature of the "spiral nebulae" was determined in *The Discovery of Our Galaxy* (Knopf, 1971). The development of astronomical ideas in their wider scientific context is discussed in a succinct survey, *A Short History of Scientific Ideas to 1900,* by Charles Singer (Oxford, 1959).

Biographies and memoirs this book drew upon include George Gamow's delightful *My World Line* (Viking, 1970), Harlow Shapley's *Through Rugged Ways to the Stars* (Scribner's, 1969), and Werner Heisenberg's *Physics and Beyond* (Harper & Row, 1971). Also *Explorer of the Universe: A Biography of George Ellery Hale,* by Helen Wright (Dutton, 1966), *Niels Bohr: The Man, His Science, and the World They Changed,* by Ruth Moore (Knopf, 1966), *Arthur Stanley Eddington,* by Clive Kilmister (Oxford, 1966) and *Life of Arthur Stanley Eddington,* by A. V. Douglas (Nelson, 1957).

Two books that tell how the 200-inch telescope on Mt. Palomar was built are *The Glass Giant of Palomar,* by David Woodbury (Dodd, Mead, 1939) and *Palomar: The World's Largest Telescope,* by Helen Wright (Macmillan, 1952).

TEXTBOOKS

Inexpensively priced astronomy textbooks can be found in used book shops, and it is a good idea to have one or two on hand when questions arise. The battered editions on my shelf include *Exploration of the Universe*, by George Abell (Holt, Rinehart and Winston, 1964), *Essentials of Astronomy*, by Lloyd Motz and Anneta Duveen (Wadsworth, 1967), *The Astronomical Universe*, by Wasley Krogdahl (Macmillan, 1962), *Introduction to Astronomy*, by Dean McLaughlin (Houghton Mifflin, 1961), and *An Introduction to Astronomy*, by Robert Baker and Laurence Frederick (Van Nostrand, 1968).

BASIC ASTRONOMY

Introductions written by professional astronomers include *Astronomy*, by Fred Hoyle (Doubleday, 1962), *Astronomy*, by Donald Menzel (Random House, 1970), and *The Universe*, by Otto Struve (MIT, 1962). A small library of books on astronomy have been produced by two popular writers, Patrick Moore in England and Isaac Asimov in the United States; both writers are prolific—Asimov astonishingly so—and rather than attempt to list their books I simply recommend them in general.

Fairly recent developments are discussed in Nigel Calder's *Violent Universe: An Eyewitness Account of the New Astronomy* (Viking, 1969) and John Glasby's *Boundaries of the Universe* (Harvard, 1971), the latter of which is the more technical.

Some of the most beautiful popular writing on astronomy was produced by Arthur Stanley Eddington and his colleague and sometime rival James Jeans. The books are somewhat outdated but still very much worth reading; try Eddington's *The Expanding Universe* (Macmillan, 1933) and *Stars and Atoms* (Yale, 1927), Jeans' *The Universe Around Us* (Cambridge, 1929) and *The Mysterious Universe* (Cambridge, 1930).

Paul Couderc's *The Wider Universe* (Arrow Books, 1960) focuses on extragalactic astronomy, while D. W. Sciama's *The Unity of the Universe* (Doubleday, 1959) touches on philosophical matters, notably Mach's principle; both are straightforward and free from technical jargon.

Two introductions filled with pictures are the *Larousse Encyclopedia of Astronomy* (Prometheus, 1959), and *Pictorial Astronomy,* by Alter, Cleminshaw and Phillips (Crowell, 1969).

The Hubble Atlas of Galaxies, by Allan Sandage (Carnegie Institution, 1961), with 50 pages of photographs of galaxies, is an inspiring book.

ADVANCED, SEMITECHNICAL AND SPECIALIZED BOOKS

These need not be forbidding to the general reader, but most require some familiarity with astronomy.

Thirty-three articles on astronomy, ranging from Earth to the quasars, are collected in *Frontiers of Astronomy* (Scientific American/Freeman, 1970). *Beyond the Milky Way* (Macmillan Sky and Telescope Library of Astronomy, Vol. 8, 1969) is slightly more technical but equally engrossing. Two other collections of interest are *Exploring the Universe,* Louise Young, editor (McGraw-Hill, 1963), and *Galaxies and the Universe,* L. Woltjer, editor (Columbia, 1968).

Radio astronomy and the astronomies of other parts of the electromagnetic spectrum beyond visible light have developed almost too rapidly for books to keep pace, but several works do provide help in understanding the non-visual astronomies: J. S. Hey, *The Radio Universe* (Oxford, 1971), J. H. Piddington, *Radio Astronomy* (Harper, 1961), H. P. Palmer and R. D. Davies, *Radio Studies of the Universe* (Routledge and Kegan Paul, 1959), John Kraus, *Radio Astronomy* (McGraw-Hill, 1966), A. G. Pacholczyk, *Radio Astrophysics* (Freeman, 1970), I. S. Shklovsky, *Cosmic Radio Waves* (Harvard, 1960), H. P. Palmer and R. D. Davies, editors, *Radio Astronomy Today* (Harvard, 1963), Colin Ronan, *Invisible Astronomy* (Lippincott, 1972), and James Ring, *Infra-red Astronomy* (University of London, 1968).

For information on quasars, see *Quasi-Stellar Objects,* by Geoffrey and Margaret Burbidge (Freeman, 1967), *Quasars,* by F. D. Kahn (Harvard, 1968), *Galaxies, Nuclei, and Quasars,* by Fred Hoyle (Harper & Row, 1965), and an account of the Halton Arp-John Bahcall debate over quasars, *The Redshift Controversy,* George Field, editor (Benjamin, 1973).

Several of the leading figures portrayed in this book left accounts of their work. Edwin Hubble's *The Realm of the Nebulae* (Yale, 1936) is a classic that demands no expertise on the part of the reader—but remember that Hubble's figures for such quantities as the distances between galaxies have since been revised upward to a great degree. *The Observational Approach to Cosmology* (Oxford, 1937) is another important Hubble volume. Otto Struve's *Stellar Evolution* (Princeton, 1950), more technical, provides an account of the lifetimes of stars written by one of those who contributed most to understanding them. From Walter Baade, who seldom lectured or published, we have *Evolution of Stars and Galaxies* (Harvard, 1963), a transcript of his 1958 Harvard Observatory lectures.

RELATIVITY

Of the biographies of Einstein, Philipp Frank's *Einstein: His Life and Times* (Knopf, 1970) deals most clearly with relativity, and is the book I would recommend to readers who want to learn equally about the man and his work. Ronald Clark's *Einstein: The Life and Times* (World, 1971) is longer and a good source of anecdotal material. Two readable shorter accounts are *Albert Einstein: Creator and Rebel*, by Banesh Hoffman (Viking, 1972), which like Frank's volume contains many arresting photographs, and *Einstein*, by Jeremy Bernstein (Viking, 1973).

The enduring public fascination with relativity has spawned many popular explanations of the theory. Perhaps the most accessible are *The ABC of Relativity*, by Bertrand Russell (George Allen, 1958), *The Universe and Dr. Einstein*, by Lincoln Barnett (Harper, 1948), and *Relativity for the Layman*, by James Coleman (Penguin, 1959), followed closely by Einstein's own *Relativity, The Special and General Theory* (Holt, 1920), and *Relativity and Common Sense*, by Hermann Bondi (Anchor, 1962).

Midway between the nontechnical and the technical levels stand works by two of Einstein's colleagues: *Four Lectures on Relativity and Space*, by Charles Proteus Steinmetz (McGraw-Hill, 1923), and *Einstein's Theory of Relativity*, by Max Born (Dover, 1965).

Gravitation, by Charles Misner, Kip Thorne, and John Wheeler (Freeman, 1973), is a massive textbook intended for graduate students in physics, but anyone interested in relativity should browse through this remarkable book, if only to have a look at its many biographical sketches, quotations, and diagrams.

Readers interested in contemporary challenges to relativity theory should consult two books by Robert Dicke, *Gravitation and the Universe* (American Philosophical Society, 1970) and *The Theoretical Significance of Experimental Relativity* (Gordon & Breach, 1964). The former is the less demanding.

Relativity is placed in the context of the physics of its day by A. D'Abro in his *The Rise of the New Physics* (Van Nostrand, 1939; Dover, 1951), and analyzed in terms of psychological context by Lewis Feuer in *Einstein and the Generations of Science* (Basic Books, 1974).

COSMOLOGY

Cosmology is a highly theoretical subject and the nonmathematical reader is best advised to read several works to gain a proper perspective. On the general level we have Jagjit Singh's rewarding *Great Ideas and Theories of Modern Cosmology* (Dover, 1961), and a charming book by F. P. Dickson, *The Bowl of Night: The Physical Universe and Scientific Thought* (MIT, 1968). The perspective of an advocate of the Big Bang theory is supplied in nonscientific terms by George Gamow's *The Creation of the Universe* (Viking, 1952), that of the Steady State in Fred Hoyle's *The Nature of the Universe* (Oxford, 1960). *Theories of the Universe,* edited by Milton Munitz (Macmillan, 1957) offers excerpts from the writings of cosmologists from Lucretius through De Sitter, Lemaître, Milne, Gamow, Bondi and Hoyle. *The Universe and Its Origin,* by H. Messel and S. T. Butler (W. & J. Mackay, 1964) offers papers by several contemporary cosmologists, as do *The Emerging Universe,*

William Saslaw, editor (Virginia, 1972), and *Cosmology, Fusion & Other Matters,* Frederick Reines, editor (Colorado, 1972), a memorial volume issued in tribute to Gamow.

Hermann Bondi's *Cosmology* (Cambridge, 1960) is a standard reference work, and semitechnical. Other helpful books on this level include *Fact and Theory in Cosmology,* by G. C. McVittie, *The Origin and Evolution of the Universe,* by Evry Schatzman (Basic Books, 1965) and *The Structure and Evolution of the Universe,* by G. J. Whitrow (Hutchinson, 1959).

J. D. North's *The Measure of the Universe* (Oxford, 1965), an extensive survey of modern cosmology, is recommended to advanced students. For cosmological writing by leading theorists themselves, see Bondi's *Rival Theories of Cosmology* (Oxford, 1960), James Jeans' *Astronomy and Cosmogony* (Dover, 1961), De Sitter's *Kosmos* (Harvard, 1932), Hubble's *The Observational Approach to Cosmology* (Oxford, 1937), Hannes Alfvén's *Worlds-Antiworlds* (Freeman, 1966), Georges Lemaître's *The Primeval Atom* (Van Nostrand, 1950), and P. J. E. Peebles' *Physical Cosmology* (Princeton, 1971). Those interested in Immanuel Kant's contributions to cosmology should note that a translation of his *Universal Natural History and Theory of the Heavens* has been published by the University of Michigan Press in unabridged form.

PHILOSOPHY OF NATURE

These are some of the sources drawn upon for the philosophical issues raised in the final chapter:

Bohr, Niels, *Atomic Theory and the Description of Nature* (Cambridge, 1934).

Bondi, Hermann, *Assumption and Myth in Modern Theories of Science* (Cambridge, 1967).

Burtt, E. A., *Metaphysical Foundations of Modern Physical Science* (Harcourt, 1927).

Carnap, Rudolf, *Philosophical Foundations of Physics: An Introduction to the Philosophy of Science* (Basic Books, 1966).

D'Abro, A., *The Evolution of Scientific Thought From Newton to Einstein* (Dover, 1950).

Eddington, Arthur Stanley, *Fundamental Theory* (Cambridge, 1946).

———, *The Philosophy of Physical Science* (Cambridge, 1939).

Gold, Thomas (ed.), with Schumacher, D. L., *The Nature of Time* (Cornell, 1963).

Heisenberg, Werner, *Across the Frontiers* (Harper & Row, 1974).

———, *Physics and Philosophy* (Harper, 1958).

Hubble, Edwin, *The Nature of Science and Other Lectures* (Huntington Library, 1954).

Jeans, James, *Physics and Philosophy* (Ann Arbor, 1958).

Planck, Max, *Scientific Autobiography and Other Papers* (Philosophical Library, 1949).

———, *The New Science* (Meridian Books, 1959).

Poincaré, Henri, *The Foundations of Science* (Science Press, 1913).

Reichenbach, Hans, *Modern Philosophy of Science* (Humanities Press, 1959).

———, *The Philosophy of Space and Time* (Dover, 1957).

Schlegel, Richard, *Time and the Physical World* (Dover, 1968).

Schilpp, Paul (ed.), *Albert Einstein: Philosopher Scientist* (Library of Living Philosophers, 1949).

Toulmin, Stephen (ed.), *Physical Reality* (Harper & Row, 1970).

Weyl, Hermann, *Philosophy of Mathematics and Natural Science* (Princeton, 1949).

Whitrow, G. J., *Natural Philosophy of Time* (Nelson, 1961).
Whitehead, Alfred North, *Essays in Science and Philosophy* (Philosophical Library, 1947).
———, *Modes of Thought* (Macmillan, 1938).
———, *Science and the Modern World* (Macmillan, 1925).

JOURNALS AND PERIODICALS

Scientific American, a leading journal of science for general readers, regularly publishes articles on astronomy, astrophysics and cosmology. Two other excellent magazines, somewhat more rigorous, are the American *Science* and the British *Nature.*

Sky and Telescope has long been the leading popular astronomy magazine in the United States, though several new publications have sprung up to meet growing interest in the subject; one should watch the newsstands. Two publications aimed at both amateur and professional astronomers are the *Bulletin of the American Astronomical Society* and *Publications of the Astronomical Society of the Pacific.* In Britain, see *Monthly Notices of the Royal Astronomical Society.*

American papers on astrophysics and extragalactic astronomy appear in the *Physical Review* and the *Astrophysical Journal;* both journals are as demanding as they are rewarding.

INDEX

279